9781022515364

Daedalus
Dialogues

DÆDALUS

Journal of the American Academy of Arts and Sciences

Contents

v   *Preface*

888   *New Trends in History*

978   *The Languages of the Humanistic Studies*

1030   *Governance of the Universities I*

1092   *Governance of the Universities II*

1157   *Higher Education in Industrial Societies*

1225   *Conference Participants*

1228   *Notes on Contributors*

1233   *Index of Authors for Dædalus, Volume 98, 1969*

1244   *Recent Issues of Dædalus*

Fall 1969:   DÆDALUS DIALOGUES

Issued as Vol. 98, No. 4, of the Proceedings of the American Academy of Arts and Sciences

Preface to the Issue "Dialogues"

THERE IS constant talk today about the need for "dialogue." The term is, in most respects, ambiguous. It may mean almost anything —the establishment of new forums to encourage the exchange of opinion; the development of new political instrumentalities, even in non-political institutions, to alter the distribution of power; the creation of new communities, with a promise of their being guided by higher ethical standards. Each of these prospects will seem compelling to the individual who is persuaded of their need; each will seem unnecessary (or even dangerous) to those who are satisfied with existing forms. When, as in the present climate, strident and angry voices are raised on every side, it is difficult to know what the intent of any radical proposal is, or whether any deserves to be taken seriously. New terms like "participation" become clichés almost as soon as they are uttered; old terms like "authority" are simply vulgarized. Rhetoric and passion take the place of common sense and reason, and we no longer know whether we are seeking new forms of representation, new rules of political procedure, new devices for airing grievances, or new ways of living together. The art of debate is lost; the convention of discussion is threatened; and rational discourse becomes the casualty of intemperate demagoguery.

The mass media (television, press, and radio) do their best to record and reproduce these "happenings" for audiences whose taste for such news would appear to be insatiable. Another generation, chancing on this "record" and having access to no other, might well conclude that men in this time were indeed mad, made so either by the gravity of their problems or by their obvious incapacity to cope with them. It is precisely because the mass media are so addicted to the sensational demanded by the consuming public that there is a special obligation to tell of events which suggest another temperament in our society, one no less concerned with truth, social justice, and moral improvement, but one somewhat more restrained in its expression. Ours is indeed a time of "dialogue," but not only of the kind that is angry and agitated. There are today unequaled opportunities for the exchange of opinion. If, as many of us believe, these opportunities must in fact be strengthened and multiplied, there is good reason to look closely at what is presently happening in certain of those places where discussion still thrives, where differences may still be expressed without offense being given. The American Academy of Arts and Sciences offers hospitality to such possibilities.

Some will say that Academy discussions, as reproduced here, are too academic and too tame; that they do not sufficiently touch on the more fundamental grievances of our time. This situation may indeed obtain, though there is reason to think that Academy conferences in fact reflect diverse opinion to an extent that is not at all common when the decibel is higher or when protestations of concern are more insistent. There are no rules of debate in Academy conferences; no principles of representation; no pretense to preparing a record that will go straight to persons in the highest places. A more modest ambition insinuates itself, creating the civility and tolerance that is so essential to good talk. The conferees are interested in one another's thought, expecting to derive pleasure and profit from it; they share with the Editors the hope that out of good talk may come a better publication that will in time reach an interested public and have its influence there. Academy conferences are intimately tied to plans for publication, in *Dædalus* or elsewhere. These are indeed "dialogues," but in the best sense; they express appreciation for the spoken (and the written) word.

Those who participated in these American Academy discussions understood that their remarks were being transcribed, though it was not expected that these comments would then be circulated widely. Our practice in the past has been to mimeograph conference transcripts, making them available to small numbers and using them to assist in planning publications. The warm and enthusiastic reception of these hastily prepared transcripts—and the constantly repeated suggestion that they be made available to larger numbers—led us to decide to experiment with their publication in this issue of *Dædalus*. Participants have been asked to review their contributions; some have made minor stylistic changes; the greatest number, however, have permitted the record to stand as originally transcribed. These are, then, the spontaneous and unrehearsed comments of men and women, seated around a table, speaking for themselves to others who were in precisely the same situation. They did not attend as representatives of institutions or professions; they did not imagine that their words carried weight except as they might persuade some individual

Preface

in the room who found them challenging. The Editors of *Dædalus* feel a deep gratitude to those who have agreed to have their contributions made available in this form.

A word needs to be said about the choice of transcripts. The Academy holds several dozen conferences each year; it would have been easy to select materials only from those transcripts having to do with the critical political and social issues of our time. A deliberate decision was made not to concentrate on these alone. Two reasons guided the Editors. To deal with such matters only would distort the picture of what the Academy was in fact doing; more importantly, perhaps, it would distort the picture of what the intellectual and academic community is seeking to accomplish. There is much to be said in support of the contemporary "political" activity of academics; it must not, however, be thought to exhaust their contributions to contemporary society.

The issue, appropriately enough, opens with selections from a conference on New Trends in History. The meeting, held in Princeton at the Institute for Advanced Study, was chaired by Professor Felix Gilbert. The ostensible purpose was to plan an issue of *Dædalus;* the larger aim was to consider how history—one of the more "traditional" university subjects—was being transformed by new scholarly interests and twentieth-century research techniques. Particularly at this time, when anti-historical bias is prevalent in so many places, it may be profitable to consider what the professional historian is doing. A great debt is owed Professor Gilbert for giving the intellectual leadership to this enterprise. Thanks are due also to Dr. Carl Kaysen, Director of the Institute, who joined the Academy in sponsoring this "planning" session.

A very different "planning" meeting took place last year at the University of California in La Jolla. This conference, despite its somewhat forbidding title, "The Languages of the Humanistic Studies," sought to develop a better understanding of the theoretical concepts and constructs that may be said to underlie humanistic endeavor. The conference was unique in many respects, not least in the professional interests of certain of its members. Although the overwhelming majority came from disciplines that would be classified generally as humanistic, an important and articulate minority came from the social and the natural sciences. Because no single part of the transcript seemed to capture the quality of the discussion, we have put together selections from certain members of the group. The selections, as will be noted quickly, come from an anthropologist, a sociologist, a philosopher, an oceanographer, who is also an expert on population studies, and a historian. It is for the reader to say whether there is an advantage in having such a wide representation of disciplines. If there is, would it not be equally important to have humanists present in a meeting where the theoretical basis of scientific and social scientific endeavor was being discussed? The Editors would say yes, unreservedly. Thanks are due to Professor Morton Bloomfield of Harvard University and Professor Roy Harvey Pearce of the University of California, La Jolla, who inspired this effort.

For both the La Jolla and the Princeton meetings, the Academy would wish to acknowledge the generous support given by the Ford Foundation. A grant from that Foundation has made possible meetings having to do with the humanities that could not otherwise have taken place.

The American Academy's interest in the problems of higher education is of long standing. On two occasions in the last sixteen months men and women from all the major constituencies of universities—trustees, administrators, students, faculty, alumni, legislators—have gathered to discuss the problems of university goals and governance. These meetings took place at the House of the Academy in Boston; selections from the conference transcripts suggest how advantageous it is to bring together informed persons from many different kinds of institutions and, also, to give thought to including some from outside the university. A great debt is owed to the Danforth Foundation for making these meetings possible. At least two issues of *Dædalus*, planned for the future, will be based on insights developed in these meetings.

Finally, we publish the first sessions of what is to be a continuing seminar on the problems of higher education in industrial societies. These sessions took place at the House of the Academy in Boston. A second meeting was held in Paris in June; there was no time to prepare this European transcript for publication. Again, the object was to bring together people who do not normally meet one another to discuss problems that have great concern for all of them. There can be no question about the importance of such international debate, carried on in an environment where each seeks to liberate himself (if only for a few days) from his more parochial concerns. The Ford Foundation is to be thanked for making this "continuing seminar" possible.

S. R. G.

New Trends in Histo

A few words are in order about the conference transcript that follows. First, it should be noted that this was a "planning" session to determine whether an issue of Dædalus on New Trends in History should be undertaken. Because the meeting was brief—only two and a half days—and there was no certainty that we would in fact pursue the theme toward publication, there was some reluctance to invite persons from abroad. All participants, therefore, came from the United States or Canada. Note, however, should be taken of the range of interests among the historians invited and of the concerns of those who came from other fields entirely—political science, economics, sociology. Next summer when members of this original group meet in Europe with historians from the United Kingdom, France, Germany, and Italy, the discussions may take on a quite different cast. If the net were spread even wider—to include scholars from the East European countries, the Soviet Union, and Japan— there would be other possibilities. It is not impossible that we will want to achieve this in a third conference —not yet planned, but that may well build on other activities in which the Academy is already involved.

Participants
Samuel H. Beer
Lee Benson
Felix Gilbert
Stephen R. Graubard
David J. Herlihy
Stanley Hoffmann
Carl Kaysen
Leonard Krieger
Thomas S. Kuhn
David Landes
Joseph Levenson
Frank E. Manuel
J. G. A. Pocock
David J. Rothman
Carl E. Schorske
Lawrence Stone
Charles Tilly

GILBERT

Some days ago I received a questionnaire I am sure you all got. One of the questions asked was: What nonhistorical scholarly periodicals are you reading? This question might yield interesting information on the general education of the historian, but I would have said that it would have been much more interesting to ask historians what scholarly *historical* periodicals they were reading. There has been a shift of interest, a shift of what historians consider to be their main task. It seems to me that we might look at the present trends in history in terms of this shift in interest and try to determine what the main concerns of the historian are now.

BENSON

Let me state a rather extreme position. In my judgment, history has become a relatively trivial enterprise. I cannot think of any significant contribution that specialists of past human behavior have made to social thought in the last fifty years. During the nineteenth century, the understanding of human behavior was dominated by the historical approach. What we think of as social science is fundamentally historical, as were the nineteenth-century social scientists. It hardly needs arguing that Marx and Weber were basically historical in their approach. I think it reasonable to claim that no historian writing during the last fifty years has significantly contributed to our thinking about the nature of human beings, the way human beings function, or how the study of human behavior can help men achieve desired social goals. If we accept that proposition, the conclusion follows that soon after the turn of the century history as a discipline began to lose intellectual vitality. As actuality, of course, history possesses unity. Everything that happens affects in some infinitesimal way everything else. We have been trying to study the past that way; that is what I think Charles Tilly means by the term "integralist" in his memorandum. It is a kind of *Gestalt*—everything is related to everything else and you cannot understand anything unless you understand everything and see the total configuration. Separating something out from the totality necessarily distorts and destroys this *Gestalt*. Having adopted this integralist approach, historians cannot contribute much to social theory.

Tilly and I were on a program together last fall on more or less the same subject; he argued that history is too important to be left to historians. The studies of political development make it perfectly clear that it is too hard to do anything significant without adopting a historical approach. The Committee on Comparative Politics of the Social Science Research Council has come to the conclusion that it has gone as far as it can with contemporary material and that it cannot develop any theories without going into historical materials. Consequently, it has recently set up a subcommittee on historical research. Tilly's key criticism is, I think, that we have failed to understand that over the long run the application to historical materials of social scientific styles of analysis, explicit conceptualization, model-building, systematic comparison, and common criteria of measurement would profoundly affect historical practice.

Historical practice cannot do simultaneously all the things that historians have tried to do. We have, in fact, been operating at an underdeveloped, structurally undifferentiated level, simultaneously trying to be philosophers and policy scientists. These are not the same things, and they require different kinds of skills and different kinds of personalities. To develop some rational division of labor would require a radical reorganization of graduate and undergraduate training and a reconceptualization of what constitutes good historical work. As of now, we have a mixed bag of things. One simply does not write a beautiful exposition in the fashion of Francis Parkman and do what Charles Beard tried to do. These tasks are quite different. If we use a single standard to judge historical work, we are going to get an undifferentiated mass that will become increasingly trivial.

If we want to practice history as identity, then it makes sense to give a survey course in American history, one in German history, and another in English history. We would then organize the major historical journals along national lines. But this organization makes no sense if we are interested in history as "policy science," in studying the past to develop more powerful theories about behavior. I regard history as "policy science" as a subdivision, a field of specialization, within the general discipline. History as policy science attempts to understand the past in order to control the present and the future. I maintain that that goal has always attracted historians, but it has been considered in conjunction with the other goals of history as philosophy, as identity, and as entertainment. Until we advance beyond that undifferentiated state, we are doomed to triviality.

MANUEL

I am moved by the idea that certain forms of the old history have to go, but for totally different reasons. In preparation for this conference, I read six or seven historical journals from various parts of the world. I read three issues of each because that is the number they gave me at the New York Public Library. So in a sense my sample was objectively chosen. I had someone look at the Russian journals for me. It was my overwhelming impression that almost a third of the articles was political history of the old-fashioned kind and could have been published in the *Historische Zeitschrift* of the late-nineteenth century. (The figure I came out with was 29.3 per cent.) This particular form of political history is still the controlling reality. Outside a very few universities, for example, at doctoral examinations they almost always ask what went on in Vienna at a certain date I don't remember. It is against this background that we must ask the question of what we should do to revitalize history-writing. The reaction I have to Lee Benson's memorandum, however, is that it represents overkill. He uses such words as "the savage internecine warfare currently distressing so many disciplines," but I do not find this much lively discussion among so many disciplines.

The one discipline outside history that I know something of is psychology. I attend their seminars, and their discussions do not have a great degree of violence. The psychologists have the same kind of compartmentalization that we do. The nice traditionalist guys are in control, and a few people are trying to make some breakthroughs. The breakthrough of, say, depth psychology in America is only a matter of thirty years of teaching. It hardly existed in the universities, even though the books did.

I think we need to provoke variety and, in the spirit of the age, a little anarchy in the historical profession; I want to see new things happening. But before Mr. Benson finishes consigning the old history, he immediately gives us more compartmentalizations. Whether they are good or bad is irrelevant. I think it foregone that they do not affect the essential problem. I would be all for opening up, but opening up to new styles. Let there be opportunities for new styles of history-thinking and history-writing to develop.

I was cautious when the Academy sponsored a study group on psycho-historical process. I learned a great deal from it, but was troubled that its first manifestoes maintained that this new history was going to be a replacement for all others. Why must there be a replacement? Why cannot different people form different configurations? The visual arts are quite creative at the present moment. I have just gone through an exhibition of sixty years of French painting, and the artists have dared to use different styles. Now why not devote ourselves to negation of the stereotyped mono-style of history-writing and open up new questions, new methods, without pretending to make one new history? This variety would be the new history. The newness in the new history that I would want to emphasize would be its openness and its pluralism.

Tilly, for example, wants to pursue a specific kind of style using quantitative methods, and he does it very well. He deals with the

French working classes in a way very different from mine, using precisely the same materials. There would be no possibility of a meeting ground between us. Tilly is quantifying the French strikes, where I would want to dramatize the personalities of the strikers and see them in terms of the universal values of the nineteenth century. There is no meeting ground, but why cannot both coexist? I surely am bored with the old-fashioned political history, and being boring is for me a moral category. It bores the kids too. I would make the break when a historical form gets to be repetitive, grinding the old stuff over and over again. I hope that the new trend in history will be pluralist, allowing for different people doing different kinds of things in different ways.

STONE

I would like to say I entirely agree, except on one fundamental point—Frank Manuel's comment that when Tilly analyzes French strikes in a quantitative way and comes up with one thing, and I analyze in an intellectual way and come up with something quite different, we go past each other and have nothing to say to each other. If this is true, I think it is tragic. It is appalling if two men studying a single phenomenon just walk past each other and have nothing to say to each other.

ROTHMAN

Policy science is almost a pejorative term. But if we are interested, as I am, in questions of society—its cohesion, its lack of cohesion—this interest ought to have pertinence to the present. If I go back and attempt to deal with society as a system that does or does not hold together, my initial encounter is one of incredible frustration. The material, the data, I would like to have before me, I do not have before me.

In a narrow sense, it is difficult to disagree with Mr. Manuel's plea for differing styles that do not necessarily have to add up. But if we are going to deal with society in any way that enables us to say something about a process, that has some relevance for our understanding of what goes on about us, I am afraid we may not be able to afford the luxury of not adding up.

If we ask about, say, riots and ghettos, we sense that the traditional restraints have broken down. A historian then asks how traditional restraints have broken down in the past and what that means. What types of restraints in other eras maintained a certain stability so that rioting was not a common behavior pattern? That is a very interesting question. But were I to try to answer it, I would need immediately all sorts of information—on what the churches were doing, on what the family was doing, on what a whole series of institutions not yet explored in anything approximating sophistication were doing. I agree fully that political history dominates the American discipline as it dominates the European. But when we are so helpless before questions that have relevance to the present, perhaps the luxury of style is less important, and some attempt should be made at a coming together to make this business cumulative so that one can say something about a society.

POCOCK

I wanted to speak up against armadillos. You may remember that fable of Kipling's in which the porcupine and the tortoise jointly turned themselves into an armadillo. This happens to be a sad story because the armadillo is a dull animal and the dialogue between the porcupine and the tortoise was delightful all the way through.

Professor Stone is troubled that two historians, doing the same thing but in ways so different that they are using different frameworks, may not have much to say to each other. Why is this situation tragic? If we assume that there is no contact and absolutely no communication between them, then that is certainly rather sad. It does not follow, however, that they must instantly look for a way of turning themselves into an armadillo, of synthesizing themselves into a different sort of animal or historian. It may quite simply happen that they read one another's works and occasionally a spark is knocked off. One admires the other's writing and something interesting may happen that means more to the man who read the book than it did to the man who wrote it. This is a normal communications system as I understand it. As long as that happens, and nothing has been said to indicate that it does not happen, then nothing particularly tragic has occurred. Manuel and Tilly are clearly in need of each other. This also opens up a huge new field.

We are in agreement with Professor Benson that historians do different things, although I am not at all happy about his attempt to categorize the things that they do. Some people are interested in exploring a particular aspect of a particular event or personality; others are interested in integrating different techniques to form something socially relevant.

If you start with the historian in a relatively early stage of his being, he is looking at all sorts of things in the past—they interest him for different reasons, he looks at them in different ways, and he has different aims. To some extent, therefore, he is practicing a congeries of subdisciplines, but he does have a way of putting these subdisciplines together to make narratives. There then occurs a technological explosion. Certain subdisciplines and subtechniques take off, if I may use that phrase. They are developed to the point where they can be the basis of different industries and spawn different internecine feuds. It then becomes possible to think about integrating these different technologies into some kind of hypertechnology or social science. But this by no means seals off the social historian who holds back from being captured by the technological explosion. The non-technological historians, not the antiquarians Tilly speaks of, finally reassert themselves and say they are going to go on writing in a way they consider valuable for the whole hierarchy. You cannot stop them from concentrating on the aesthetic, if that is their preference, and thus the savage internecine warfare would leave them, as I hope it would leave me, totally untouched.

KUHN

Frank Manuel's parable of his passing Tilly in the night is still doing us a certain amount of harm, and I would like to reinforce what Professor Pocock has said. Nobody has mentioned as yet the most likely way in which these apparently quite different approaches come together—and that is when a man in the next generation writes a book on the strikes in France. His work will clearly show the effect of having read both of these earlier works. The confrontation does not have to happen between the two people who are initially responsible for developing the two approaches. In this way—which you may or may not want to call cumulative—the various approaches get picked up. They influence future work even where there is perhaps only a moderate amount of direct communication between the original authors. I consider communication through third parties to be very important—potentially and actually—in the development of a discipline of this sort.

I am somewhat distressed about another point: the extent to which we may already be accepting without question the notion of history as a policy science, or the idea that policy science is a form that comes naturally to history and one we should strive to develop further. I would therefore like to introduce a distinction, which is certainly never going to be very clear cut, between utilizing data from the past in order to do social science and utilizing them in order to construct history. The same data can, it seems to me, be utilized in either way. Assuming that someone is being a historian simply because he deals with past data will run us all sorts of trouble. He may be using them to do social science.

I am not at all sure that the deliberate search for generalizations out of historical data is going to produce the kinds of generalizations that will have direct and immediate applicability to certain of the more obvious contemporary problems. One should keep trying, but I should like to see the attempt left to the social scientist, not the historian. I suppose that where one has got from history some sense of how to act in a particular pressing contemporary situation, one has got it not through generalizations produced from history, but from exposure to concrete historical narratives which increase one's experience of concrete events, very much in the absence of generalizations that one could apply directly to the situation. I am worried by the extent to which the attempt to turn historical narrative into a device for discovering generalizations with contemporary applicability may deprive the historian's enterprise of certain of the things that are most uniquely its own and have been perhaps its most fruitful products.

GILBERT

Until the nineteenth century, history was never anything but a congeries. It was political history; it was sometimes economic history; it was a part of moral philosophy. The basis of nineteenth-century history was the nation-state; people believed that politics was the decisive factor and that everything could be grouped around it. Politics provided the unity of history at that time.

I have always believed that every historian should do what he wants to do. Nevertheless, if you take the nineteenth century and the early-twentieth century, then even scholars who worked in entirely different fields had the same background color—politics. They might go in different directions, but they still maintained continuity or connection among one another and among their varied works. A question which we must face is whether at the present time we have anything of the same importance as politics. One of the topics of the present meeting might be whether social history can fulfill this function.

I come now finally to my last comment and that is on history as a policy science. This may be a false problem. The issue seems to me to have reached undue significance in this country because in the United States more than in any other country history is identified as being history of the United States. In no European country would you identify to the same extent the entire past with national history.

LANDES

But no other country devotes so much of its teaching resources and its time in history to the study of the history of other nations as the United States does. All you have to do is look at the curriculum.

GILBERT

It is not a question of how much time you devote, but of the extent to which the historians are concerned with the history of their own country and consider this history as their only problem. This is much more true of the United States than of Europe.

SCHORSKE

Felix Gilbert was saying that history is studied and taught essentially in a national frame in this country more than in Europe, not that people are devoting themselves to American history more here than they are to German history in Germany or to French history in France. The group meeting here is probably atypical in that none of us would classify ourselves primarily as national historians. In the country as a whole, however, there is a French historical society, a German one, and a this and a that. Our curricula, historical magazine articles, and the like are built on this principle.

LANDES

Aren't we kidding ourselves in one respect? It would be hard for anyone in France to classify himself as a French historian as such, because French history is treated as the larger context within which he operates. French historians are specialists in periods of French history; there are Revolutionary specialists and so forth. But our courses are also broken down accordingly. We do not just teach French history; we teach courses on the Revolution, on France since 1815, or on another period. This is misleading us into thinking that somehow in America we emphasize nations and they do not. They do work within the national context there as well.

POCOCK

In New Zealand one studies the history of New Zealand, of which admittedly there isn't much. One studies New Zealand history to get New Zealand, oneself, other New Zealanders, and still other people located somewhere in space and time, particularly in respect to one's own past.

There are clearly two issues here. One, history will obviously go on being written, studied, and taught in that way. Americans will go on studying American history for exactly the same reason. There happens to be more of it; that is the only difference. And for this reason we have to admit that history as identity, to use Professor Benson's phrase, will go on happening. It will go on being important and will affect the way all historians do their thinking. We could argue, reaching unanimity fairly soon, that there must be a better way of organizing a history curriculum than on the assumption that history consists of the pasts of various groups seeking to clarify and maintain their identity in the present. Nevertheless, history will go on being such a study, and we ourselves as individuals will to some extent go on thinking about history in that way. Thus, the alleged sterility of the national approach to history seems to me ambiguous at best.

MANUEL

Listening to this, I summarize what I have heard in two ways: One, there is an "identity crisis" that is at the present moment quite universal. One is quite in the temper of the times in proclaiming this identity crisis of history in the most exaggerated terms. But one of the ways out of an identity crisis may be to refuse to see the world in terms of identity crisis. One can always deny history's need to have *one* identity.

There are a number of people who are not obsessed with the identity crisis in the terms you are defining it. Clearly, the expressed need for this identity is historically determined. You want to be precisely like the boys in the other disciplines whose fields apparently have identities. If you look a little more deeply, you might find out that they do not have clear identities any more than you do. Perhaps this is an age where the fixed idea that a discipline ought to have an identity is at the heart of the crisis. This word "discipline" comes up again and again, but the more I hear it, the more I am convinced that it is not there.

I think that each historian should try to develop his own style, his own individual identity. In any given epoch, one can try hard to be different from others, but groupings of styles inevitably occur, so there is never anything quite so simple as "doing your own thing" in an absolute sense. This really means that groups of people will, in accordance with their physical and psychological natures, develop styles that are fitting to them. That is not anarchy.

By "style" I mean that different groups will have different emphases. I will train my students my way, in my style, and take this training seriously. I will tell people very openly that this is the style I am trying to develop. If they want to learn that style, they can come to me, but I would like to have other people to whom I can send them if they want to learn another style. This is not quite so wild in its reality as it may have sounded when I was emphasizing it earlier.

The second crisis I hear is the crisis of relevance; here the problem is discussed under headings such as "history as policy science," "history as philosophy," "history as essence," "history as entertainment." This search for a quintessential definition of history is really another way of asking: What is the relevance of the kind of history you are doing? You wake up in the morning, and you have got to go through those damn medieval charters. What the hell are you doing it for? What relevance does this work have to the morning's papers, to your whole life?

I feel certain that there are different relevances. One person will want in his historical researches a solution to the problem of cohesiveness in this society because he feels things are going on the rocks. He will argue that if a deep study were made of what held things together in the past, a kind of Parsonian history, this would be most relevant and useful in the present context. Another historian might feel most relevant when he was being apparently most irrelevant by someone else's criteria, when he was being most abstract. For example, I listened about a year ago to a Polish philosopher who thought of the practice of intellectual history in and of itself as an expression of freedom in his society. There can be various kinds of relevancy. Why must we insist that there be one? I know the relevance of my kind of historical research to me, and it is not the relevance that Mr. Rothman was talking about. An element that I would like to introduce here as desirable is the historian's need for self-consciousness in both the search for identity and the search for relevance.

STONE

I do not agree with Frank Manuel. I do think that there is an identity crisis and its explanation is quite clear. For about fifty or sixty years there was a unity of history; politics and the constitution were thought to be the central theme of history. By understanding them, you could understand everything. This has now broken down, and few people under the age of fifty believe this any more. The problem we then face is how one deals with history when there is no unified central core in which to operate.

Tilly says that we should become specialists in specific disciplines. This proposition seems to me, however, to call for precisely what our despised elders were doing. They took politics as the central theme and specialized in this; they got incredibly narrow and missed explanations. The notion that we all ought to specialize —some of us as political historians, some as demographic historians, some as social historians, and so on—might lead us toward an abyss just as bad as the one that our elders got us into.

Despite the remarks of Kuhn and others, I still think it would be terrible if Tilly and Manuel could not communicate about strikes in France. They are obviously dealing with a single phenomenon. This bloody strike happened, and certain people made it happen. If they have nothing to tell each other on this subject, I do not know why we do history. Kuhn's theory was that the reader was the only person who could get these two together. I do not think that is good enough. It will be catastrophic if we train historical demographers who ignore the history of religion. This will be just as bad as those old-fashioned political historians grinding out their stuff in a vacuum.

Finally, I should like to say a word for the national historians. It seems to me that to call oneself a national historian has become almost as pejorative as calling oneself a racist. History has to be made manageable; progress in history has always taken place by concentrating on a narrow part in great detail. To study the West since 1500, the unit one would naturally try to take is the nation-state. Obviously if you study English history, ignoring French history, you are going to get into serious trouble, but that is not the issue. We ought, it seems to me, to train people to look at societies as a whole, integrally, and resist the temptation of such drastic specialization as Tilly talks about.

BENSON

May I respond? It seems clear that at any conference there will be difficulties in communication. I do not think that we are likely to solve those difficulties in two days, and that by Saturday afternoon we will all be talking to one another. It seems useful, however, to separate out some different terms. My notion of the unity of history as actuality is quite different from my notion of the unity of history as a discipline. I do not think it is possible to do a useful study of any phenomenon by looking only at its political aspects and ignoring the others.

I would argue that the crux of the matter is the purpose for which one studies a particular phenomenon or set of phenomena. Probably no other phenomenon has had so many man-years of scholarship devoted to it as the American Civil War, because there are more American historians than any other kind. The result is trivial. If one were called before some bar to justify the incredible amount of resources and intelligence and effort expended in this field, truthfulness would force one to be silent. Nothing meaningful has emerged out of the study of the American Civil War.

POCOCK

Why not? A lot of interesting information about the American Civil War has emerged.

BENSON

But on a trivial level.

POCOCK

What do you mean by trivial?

BENSON

Trivial in the sense that it does not contribute to any generalized knowledge of human behavior. It does not enable anyone to specify the conditions under which political systems are likely to be unable to resolve differences through the normal peaceful means that the system provides. It provides no information useful to a policy adviser or decision-maker concerned with the problem of how to maintain a political system made up of subcultures. The American Civil War has enormous relevance to French Canadians, Czechs, Slovaks, Belgians, Yugoslavs, Indians, Nigerians, and all those political entities and systems which contain within a single geographic boundary people of different subcultures geographically separated. The American Civil War has not been studied in a way that would make the results at all useful to those people.

KAYSEN

Did this work fail to provide an answer to sharp questions? Could you answer the question: What made this war happen? Could you say: these elements of political structure, W per cent; these elements of individual history of the different states, X per cent; these elements of ethnic difference in the composition of the population, Y per cent; these elements of personality, Z per cent? Such a statement might be perfectly ungeneralizable in the sense that Tom Kuhn was talking about earlier, but it would say this is how I, as a historian of the Civil War, explain what happened.

BENSON

Nobody is now entitled to make that kind of statement. There is less agreement today among "specialists of American Civil War causation" than there was a long time ago. If one looks at the history of the literature on the subject, it looks like a Buddhist wheel of fate; it just goes round and round and round, making no progress whatsoever.

New Trends in History

LEVENSON

Let me bring the remarks of Lawrence Stone and Lee Benson to bear on a given subject: the question of whether or not national history has a particular validity. You might study the American Civil War in conjunction with the contemporary or almost contemporary civil war that went on in China during the 1850's and 1860's. But what would you get if you did this? On the one hand, you would uncover a certain amount of similarity which might be rather titillating and exciting for a while. You could say that, by and large, in both cases it was north versus south. You might say that the British Parliament considered what it should do about intervention. It decided for intervention in China, but against it in America. You can put together a few other little things like that, but essentially a specific study of the Taiping Rebellion would not go far in explaining the American Civil War. Far more important for explaining either the Chinese or the American Civil War is a solid contextual grounding in the histories of these countries. If you leave out those great webs, you are going to have some rather spurious and somewhat entrancing correspondence, but nothing more.

This raises the whole question of the idiosyncrasy of history. We have thrown out the old idea that historical facts are really respectable as history only if we can prove their uniqueness. But we are indeed concerned with the relevance of our studies to a lot of things outside this work. One fruitful way to get started in doing history is one that takes a middle road between an emphasis on the idiosyncratic and a leap toward homogenizing a lot of things that are really different. A historian can get fairly far along if he starts out defining his subject by seeing what is comparable between two points in time. And yet the analogy fails because there are no absolute correspondences in history. In this sense, there is always a certain something that is unique. This uniqueness need not defeat us, however; nor does it have anything to do with triviality.

Certain things are comparable in very interesting ways. Only when we measure the distance between the comparable and the analogous do we see what is relevant about the history one is studying. I would illustrate this with an example from Chinese history. The example has to do with the famous slogan "Red versus Expert" or "Red over Expert." The current Maoist line is "First Red then Expert." Some people sense analogy here where I think they should sense only comparability and then see why the analogy is not quite right. They sense analogy with the old Confucian bureaucratic assumptions that the official was not to be an expert, but was to have a commitment to a world-view and that this world-view was the most important thing about him; it validated his position, and expertise was denigrated. If you say "First Red then Expert," it sounds as though one is saying that. Thus people suggest that the old Confucian idea is still there; that the lay public may think something has happened, but this is essentially China.

The interesting aspect of Chinese history is that you can find an absolute analogy to what we have today and, therefore, presumably a way to understand the present. The historian's antennae should be out when he sees something like "Red" and "Expert." If you go in and try to solve the problem of just what exactly is *different,* you have in capsule form what has happened in China over the last hundred years—and not just the chronicle of what happened in 1873 and 1874.

SCHORSKE

I am bothered about one thing in all of Lee Benson's examples; they do not raise the one kind of question that is peculiarly historical. Historians all try to explicate a change in state from point A in time to point B in time. Every historical question addresses itself to a problem of a change in state. If that is the least common denominator for all kinds of historical inquiry, it does suggest a primacy for a kind of approach to human experience that differs from the demands that Lee makes on history. Tom Kuhn said earlier that there is a great difference between using the materials of the past for the determination or illumination of one or another set of principles, and using the past to make sense of the transformation of a given corpus of human behavior by invoking principles whose validity is never fully demonstrated in the course of their application. If a question is going to be historical, it has to focus primarily on a change in state over time. Can everybody agree with that statement?

BENSON

I do not know of *any kind* of inquiry that does not involve exactly that question. Take the Nigerian Civil War in 1967 and the American Civil War in 1861. I find it extraordinarily difficult to conceive of a historical explanation of the American Civil War and a nonhistorical explanation of the Nigerian Civil War.

KAYSEN

But one can think of lots of social processes that have nonhistorical explanations.

New Trends in History

BENSON

That don't involve change over time?

KAYSEN

Certain social processes might conceivably be explained without recourse to what Carl Schorske offers as the unique mark of historical inquiry. I am willing to offer you a bucketful of examples of such processes.

BENSON

I do not know of any study of behavior which does not involve a change in state from point A in time to point B in time.

GRAUBARD

I do think that Carl Kaysen's earlier question must be answered. However unsatisfactory we may think many of the histories of the American Civil War, certain questions seemed relevant, and even if they were not answered fully, there is available a kind of knowledge about the event that did not exist fifty years ago. This does not mean that we now agree about the causes of the war. The statement that there has been no significant contribution made in fifty years by Civil War historians is, in my view, an exaggeration. Lee Benson is arguing that the nineteenth-century historian was able to create a certain kind of synoptic history. I can only think of one or two nineteenth-century historians who in fact did what Lee is urging all historians to do. The detailed analysis of human behavior was not the ambition of many of the most prominent historians of the nineteenth century.

When historians professionalized themselves, when they became a profession and no longer a group of artists working as historians, part of the influence of the chief historical school—the Germanic one—was in the direction of making history essentially the study of politics. Today there is not this acceptance of the primacy of politics among younger historians. It is important that we not romanticize what historians once were. Rather, we should analyze what caused the diversion from an acceptance of the primacy of politics to the present situation where politics is not neglected, but is simply no longer in the undisputed lead.

ROTHMAN

In some ways, the notion of "history as policy science" is pejorative. It contains a dedication not so much to subject matter as to the notion of process. This notion of process is paramount. I do not want an immediate relevance. I do not think that my task is to help Mayor Lindsay's group put down riots in Harlem. On the other hand, after examining a particular question, I have to be able—for history's sake, for the politician's sake—to say that I have understood process.

In some ways, comparative history is a lot more acceptable to many people, but are not the innovations that comparative history is supposed to bring refinements of process? Comparative history has become more scientific in some ways because we now have controls upon our data and from this a sense of how one organizes material.

Some of the dissatisfaction with the focus on politics has been its rigidity, the absence of rigorousness in analysis. There is dissatisfaction with the notion that one can ultimately find the answers to political questions simply by staying within a political context. This frustration has led to the displacement of politics. Another focus need not be substituted for politics; nevertheless, we must open up the study of politics. Politics is not what we are after; we are looking for some notion of process. Where we look should be dictated not by predetermined boxes, but by where that material takes us.

When Carl Kaysen asked the question about Civil War causality, Lee Benson responded very properly, I think. Although Carl seemed to argue that this kind of question was a type of question that Lee might want, he phrased it very much in terms of a research design that has not yet been operative in historical studies in this field.

POCOCK

I have been trying my hand at devising models of social situations in which what we call historiography can be said to be present. The type of thing I came up with relates very much to what Stephen Graubard was saying. I see historiography as largely a political product, using the word "political" in a particular sense. I see society possessing certain data about its past, which it studies for the relevance that the past has to the present, for some aim or function that the past can carry out in the present. If society did not have this sense, it would not bother to preserve the past or to study it at all. Very simply, I want to categorize this function of the past in the present as legitimation, orientation, or explanation.

The crucial point comes when, in the course of either political or scholarly controversy or in a mixture of both, there occurs a certain breakthrough in the modes shown in the past. As a result, somebody then sees that this past is so complex a field of study that it has to be studied with relevance to itself and not to the present at all. At such a point, the historian discovers that what went on in the past did not exist merely to be relevant to the present at all. In European historiography one can date the approximate moment when this happened—about 1560 when Cujacius was asked to apply his knowledge of Roman law to a contemporary problem, and replied by asking: "What has that to do with the praetor's edict?" Cujacius could be called arid or trivial or many other harsh words, but the change he represents is of immense importance to the world-view, the intellectual life, of his society.

This is why I want us to deepen very much Mr. Benson's reference to history as identity. When I am looking at a past, I have a profound commitment to seeing it as continuous to the present. Yet I am bound also to see it as distant. Thus, I am engaging in an exercise of self-identification of a deep and subtle kind, since I am vastly enlarging the context in which my identity or the identity of the praetor's edict has to be understood. When this happens, the past is not being seen as perennial. One is looking at the uniqueness of that particular moment in the past. Perhaps as a next step, we may put that moment in a process, culminating in the present.

Some historians try to restore the exercise of seeing the past in some way that is relevant to the present; they look for new analogies, new modes of familiarity. Another—a Cujacius—will be interested in a particular moment in the past for the very complicated stimulus that comes to him from seeing the past as unique. A third will concentrate on the phenomena of process. All these activities will go on together, very often in the mind of one man. As time goes on, the number of things in a society's recoverable past that it considers relevant to its present will vastly increase. Clearly the shift from a political to a societal emphasis has come about simply because we see politics to be societal. In each and every one of these past relationships, which multiply as time goes on, the unique process may be seen as reduplicating itself. A dialogue will constantly be going on between individuals who take these three positions in respect to past relationships.

If you look at history as a social scientific function, the historian is simply an individual who functions somewhere in the context of society. He may do so in an enormous variety of ways, and we are then left to discuss, as far as I can see, only the operationally, but not philosophically, important question as to how far it is necessary and desirable that individuals conducting themselves in this complex of activities should specify and specialize in the different aspects of the activity that they are conducting.

GILBERT

I agree with Mr. Pocock's comments on the variety of attitudes of individual historians to the past and on today's interest in the societal process. While he was talking, I wondered whether this interest in the societal process has something to do with the fact that history has become an institutionalized profession. In consequence the work of the historian is meant to be useful for society and the historian feels that he ought to do something useful for society. At the moment, however, society has the feeling that history is no longer a very useful profession, that it does not fulfill any purpose. That is why Mr. Pocock's argument becomes somewhat doubtful, somewhat questionable.

POCOCK

Society merely has to decide whether or not to deprive the contemporary Cujacius of his chair. A Cujacius is in a position to make considerable demands on society, and he has got to say at some point that his work is a necessary part of humanity. To do that, he has to take a big view of human history. For reasons that cannot be specified on a functional level, he is interested in the present. He simply demands of his society an enlargement of its consciousness. If society will not pay because it does not think that this enlargement is useful, then you are left with an argument between a Cujacius and his society—an argument that has gone on now for some four centuries and one that I personally would happily see continue for four more centuries.

GILBERT

I would also continue it, but at the same time the question arises as to whether the argument must not be put on a new basis if the political center of history disappears and if a new interest in social history has developed.

KAYSEN

I am a little troubled by Mr. Pocock's use of the word "unique." It seems to me to have potentially quite different meanings. As I understood his statement, only in one sense of "unique" will the distinction that he has made hold. That sense is the dictionary definition of unique as nonrecurrent and so forth. Another sense might perhaps be most easily expressed in the following way: A particular point in a sample space is unlikely ever to be observed again. That likelihood can be made as small as you like, but you can still describe the sample space in which this is embedded. It is only in the first sense of unique that I can understand Mr. Pocock's distinction. If you accept the possibility that you can say everything that is interesting to be said about a society, a period in time, or a set of social institutions in terms of some finite collection of descriptive categories, then the edge of the distinction between the unique and the process disappears. This is perhaps a philosophical question that we should grope with, but Mr. Pocock killed it when he framed the problem in the terms that he did.

POCOCK

I did not mean that the study of what I called "unique" precluded the study of process. In fact, it seems to me on the whole to have started it off. Nor did I mean that the study of Roman law precluded the comparison of Roman law with other legal systems. Again, it may have started it off. I used the word "unique" to characterize something worth studying for its own sake. You can study a thing for its own sake and for its place in a larger pattern at the same time. These are not mutually exclusive positions, but they are distinct. Though they go on coexisting, that coexistence will involve some distinctness between them.

It seems to me that the historian is not someone who studies the past for its relevance to the present, nor is he someone who studies the past for its nonrelevance to the present. He engages in a perpetual dialectic between these two positions. This imposes plurality on the practice of history.

BEER

You can be interested in the general without necessarily being interested in the perennial. You can also be interested in the singular without being interested in the unique. I think Mr. Pocock's distinction is great when one states it abstractly, but when one comes down to actual work, such distinctions tend to become much lighter. We have no social science that deals purely with the perennial or the universal. I doubt if we will ever get one. Nor can I imagine a history that is purely singular or unique. It is interpreted in some way, implicitly or explicitly.

I feel that we are rushing by one another. The historians here are doing very well by the social sciences in this discussion. From the comments thus far, I gather certain historians are rushing toward us; but among the social scientists there are people rushing toward history for a lot of reasons.

BENSON

There has been a rush from history into social science and from social science into history because it is not possible to deal with any complex human phenomena except over long time periods. All good explanations of phenomena are historical in character. The great contributions to social thought about the behavior of human beings are historical—Marx, Comte, Tocqueville, and so forth. They focused on major problems, on large-scale social change; in effect, they were trying to study political development contemporaneously. The political scientist has now recognized that he cannot do anything worthwhile ahistorically and consequently he is rushing toward history. Activities are now regarded as being separate simply because the men who do them get their paychecks from different departments. This imposes an unreality on the fundamental reality of what we need to do in order to get at the problems that we are dealing with.

What are the significant questions that historians are dealing with? Charles Beard and Frederick Jackson Turner were not concerned about the particularity of the United States. Turner specifically said that the history of the United States is interesting chiefly because it provides a laboratory for the development of social science. If you compare John Burgess' book on the period from 1817 to 1858 and any book written now, you will see the vast shift in perspective. Burgess looked at the history of the United States between those dates in order to derive large-scale generalizations about human behavior that would enable men to understand and thereby control human behavior. I would argue that this kind of history is very different from "history as identity." We should recognize that they are different and that we need to have some kind of structural differentiation in the historical disciplines.

The only thing that some historians have in common with other men who study past human behavior in the same area of inquiry is that they have to use documents that they did not create. It seems to me that the only thing that clearly distinguishes "historians" from "nonhistorians" is the source material from which their data come. In that sense I do think there is something that all historians have in common; there are certain kinds of general problems that are related to the historian's condition—that is, he must use documents that were not created for the purposes to which he puts them.

HOFFMANN

How do you think this distinguishes a historian from anyone else in the social sciences? What do you think the social scientists work with if not documents? This distinction is most artificial. You define the social scientist as one who works exclusively on what happens today, which is not a definition I can accept. Many social scientists work with the past and use the same kinds of materials. Political scientists have been rushing toward history for the longest time. Moreover, many political scientists or social scientists have absolutely nothing to say to one another, just as certain historians have absolutely nothing to say to another group of historians. After listening to this discussion, I wonder whether the only possible distinction is that the historian teaches in a department of history and the social scientist teaches in another department. The same plurality exists in political science.

POCOCK

Do you, in fact, find a social scientist who does not give a damn about the connections between a particular phenomenon and other phenomena?

HOFFMANN

You have one sitting right here.

KRIEGER

May I leave the subject of the relations of the historian and the social scientist for a moment and direct myself to the question of what a historian now feels that is different from what a historian felt before these new trends appeared? The difference seems to me to lie in the new consciousness of the demand made upon the historian who makes any kind of general statement. A historian used to take impressionistically the general context, but he can no longer do so in good conscience. Many of the newer tendencies are attempts to build a more solid basis for the general cultural or psychological context that has always been the historian's stock in trade. When one moved at an earlier time from the description of a particular event or idea to a more general kind of knowledge—whether this be a general law of human behavior or the contribution of a particular event to a national history—the process of generalization was by and large an amateur endeavor, one that used a fairly primitive kind of logic. This form of low-level generalization would not now meet the criteria of what historians take to constitute justifiable assertions. Thus, it seems to me that the new trends tend to be along the lines of giving a more solid foundation to the more general kinds of truths a historian works with, in terms of the context for a particular event as well as of process.

SCHORSKE

When scholars address themselves to a discipline in which they are not trained, they do something different to it from what those within the discipline would do in terms of manipulating the source material. For example, Leonard Krieger's work consists in analyzing discursive prose or concepts that have been worked into a systematic order in the past and finding the places where the logic of logical men breaks down. The point at which a philosopher ceases to be consistent provides great opportunities for the intellectual historian if he asks why the logic breaks down at that particular point. The failure of the human mind in precisely that activity to which it has dedicated itself suggests some subliminal problem that cannot be handled by thought. One can then analyze the contradiction ecologically, psychologically, or by whatever principles of analysis are available and try to indicate how the social climate has entered the intellectual artifact. I do not know whether the social historians are engaged in that kind of thing, but this is certainly what I have been doing iconographically with art works. I watch for the fault lines in art, for shifts that adumbrate or clearly manifest change in the society.

GILBERT

I would like to go back to Leonard Krieger's comment that the historian has to be more scientific in his statements than he was at earlier times. I would want to ask at this point whether this is only an attempt at being more scientific or whether it also means a shift in the interest and the concepts of the historian. As soon as you begin to go into new materials don't you have a shift in concept?

In his memorandum Mr. Rothman talks about the various institutions that he is investigating—family, church, incomes, and so forth. What is your main interest? To make use of information which so far has not been adequately investigated or to establish a connection among these institutions which previously people have not seen?

ROTHMAN

In direct response, in the American case these institutions have not been heavily investigated. Secondly, if the focus becomes societal, the connections must be made. Until quite recently most institutional histories in the American instance were highly localized within the particular institution. The historian seldom asked questions about the social origins of the people, or about where the people participating in the political process actually lived. If the perspective is to be societal, there has to be unprecedented effort at integration.

I am particularly interested in the question of the cohesion or lack of cohesion of a social organization. The question of change is critical here, but to ask that kind of question one must make some effort at bringing together these diverse kinds of materials, an effort notably absent in my field.

MANUEL

I hear in this discussion two leitmotifs: One is an attempt to define similitude; the other, an attempt to define uniqueness. My problem becomes complex as I become increasingly conscious that similitude itself is a historical category. The idea of similitude was very different in the seventeenth century. There are historical mental structures that determine ideas of what is similar. This notion is obviously derived from Michel Foucault. At the present moment we are having problems because we are perhaps in a transitional stage of defining what is similar or what is unique. When you say there is difference or uniqueness, I hear from where I sit —"We have a new way of seeing change." In the seventeenth century, they would have said there is no change involved here.

Thus, in the very act of acquiring new material, I have to redefine the question of whether things are similar or unique. If they are not similar, I have to redefine what I mean by uniqueness. At this moment, different people in this room are going to have different models of similitude or uniqueness. When Felix Gilbert started the meeting, he gave me a perfect model of similitude. He jestingly put comparative and contemporary together because they both began with the letters "co." Some ages would have accepted that as a reasonable form of taxonomy; this would certainly have been true of some very serious work in the seventeenth century. To me, one is not being any more "scientific" when one introduces new models of similarity by means of mathematical quantification. Quantified data are, nevertheless, expressions of similarities of a cogent and potent kind, and I have to wrestle with them. I can easily conceive of somebody a hundred years later meaning something different by similar and unique. I cannot, however, neglect your materials; I would have to put on blinders if I did.

STONE

What, for example, do we see as similarity and change in the seventeenth century that the seventeenth century did not or would not have perceived?

MANUEL

I will give you an example straight out of Porta. If two things were physically contiguous, they would have been considered similar in certain respects.

LANDES

If you were leaning against a tree, the tree and you would be similar?

MANUEL

There would be similarity because you are both in this spot at that time. They would not say the man is the tree, but there is a sense in which the category of something being like something else will depend upon their being contiguous.

The other day I sat at a psychological conference where they got to talking about the similarity of childhood experience to other kinds of experience—for example, that of schizophrenics, or the early spatial sense in some primates. They were saying that these things are similar. I raised the question: "In what way are they similar?" The question came as quite a shock to them. The similarity was difficult to define. I do not think that categories of similitude and uniqueness are eternal; they are historical.

STONE

I concede that you have this situation in the description of natural phenomena or supernatural phenomena.

MANUEL

If I walk into a room, is it a social phenomenon? If I influence you by thinking hard in the next room, is that a social phenomenon or only a natural phenomenon? That was a common seventeenth-century belief. Do my effluvia get into you by shaking your hand? Today, we say this is charisma.

LANDES

Frank Manuel has raised an interesting problem. Granted that we look at things differently from other places and times and that even a thing like similarity turns out to be complex, would you admit or deny that over time there is some improvement in our perceptions? Would you say that human perceptions have remained roughly comparable; that it makes no difference whether or not you see effluvia; that it makes no difference whether or not you see similarity between worms and the earth because the worm is in the earth? Would you accept the possibility that there is some kind of scientific development in our perceptions?

MANUEL

This issue is fairly fundamental. Certain tribes do not see the similarity between a picture and a man. Today I go around seeing things as similar that are clearly not similar in certain respects. Maybe I am the crazy one. That is why I cannot be so sure whether the sense of similitude of a primitive tribesman or mine is superior.

ROTHMAN

You certainly cannot mean that, therefore, the notion of cumulative history raises certain intellectual difficulties to intellectual historians as a mode of procedure. Is that the tenor of your remarks?

MANUEL

The tenor is merely to increase the skepticism with respect to making a sharp distinction between similarity and dissimilarity, between uniqueness and regularity.

LEVENSON

Nevertheless, there is also the danger of going over to a radical skepticism. We are all familiar with the cautionary skepticism that tells us about the history of history. A person writing a history of his past becomes the next generation's primary source. Later historians speak knowingly of just why an earlier scholar interpreted his past in such and such a fashion and spoke for his age. People use Gibbon as a specimen not so much for the history of Rome, but for the history of the Augustan Age in England. This kind of relativism is a coin of the realm for historians; I myself feel the great importance of relativism, but only up to a certain point. One also has to have, maybe only as an article of faith, some particular stand about doing the best you can toward your own absolutes.

MANUEL

Like Hume, I abandon my absolute skepticism when I write history.

POCOCK

In my memorandum, I tried to describe a method that consists of defining political thought by exploring the language system—and I use the word "system" with considerable trepidation—which people in a political society have available to them for the purposes of political discourse. This language system under investigation is seen to have many uses that are going on simultaneously. I am particularly interested in finding through the language system the way that a political society sees its system and its politics as existing in time, the image of time that they have as a result of seeing themselves that way and being what they are. There are many theoretical problems linked to this approach. It is clearly not a technique that permits one to make many general comparative statements about the way political terms and ideas function in a society, though I do try to make a few. On the other hand, I am not committed to the history of political thought in one system at a time. I once even applied a version of this analytical process to some translations into English by various hands from various Chinese philosophers, and certain interesting observations on political ideas emerged from this work.

At the moment I am working on Renaissance political thought. I am trying to argue that Renaissance thinkers were involved in a pattern of implications stemming from their views on the relative irrationality of the particular event and, to some extent, the instability of particular political structures. This instability was conceived of in secular terms. A great deal of political thought was about secular time and how you survived in it; how you dealt with custom on the one hand and with fortune on the other. A chain of ideas emerges, beginning with Florentine civic humanism and re-occurring in a very interesting way in the quite different political and social systems of seventeenth-century England and eighteenth-century America. You have a transfer of civic humanist ideas from a city-state environment to an agrarian and, on the whole, monarchical environment.

This analytic process enables one to view a political thought by isolating what—borrowing a term from Professor Kuhn—paradigms people in the political society had for use in political discussion. It stops short to some degree of the problem which is raised once one uses a word like "ideology." I am at the moment much more interested in establishing the language system and the history of the language system than I am in trying to establish connections between that language system and other aspects of the political society, be they social, psycho-social, or psychological.

One has to get some clarity about the history of political language systems before one sets out on the further task of relating the language system to other aspects of social reality. By the way, I want to say that anyone who talks about relating theory to reality will have to be told firmly that theory is ruddy well a part of reality.

LEVENSON

What are your means of establishing the language system?

POCOCK

To begin with, you look at the words people use, and you clearly find certain new terms or certain new usages. But if you are looking at European political thought, you are of course looking at the thought of Christians conducted in the presence of a rather long-standing cultural-philosophical tradition. They are using words that come from all sorts of sources. Some will originate in the institutional vocabularies of the particular society; others will come from a cultural tradition. One has to sort them out realistically and intelligently. You have got to do a lot of abstracting to discover how political language functions, because what you are studying is, after all, an activity of abstraction.

LEVENSON

I would like to suggest another way in which close attention to language can be a method of intellectual history. Certain terms persist, but their connotations change. Something that can be defined literally may become metaphorized. Although the term may be used metaphorically, it retains a certain sense of the past, a conscious suggestion of comparison with an age that no longer persists, but is alive and vital in people's minds. One example is the famous concept in Confucian political thought of the exchange of the mandate. When one dynasty hands over the torch to another, the mandate of heaven is exchanged. The common Chinese noun for this exchange is *ko-ming*. If you look in the modern Chinese dictionary for modern contemporary vernacular Chinese, you find that *ko-ming* is the term used for revolution. This word did not always mean revolution; its connotation has been metaphorized.

You see this process of metaphorization in the thought of Sun Yat-sen, who is called the father of the Chinese republic. In 1895 or '96 he had to leave China and headed for Japan; they knew he was coming. When he landed there, a friend was with him and wrote the following account in his diary. They saw a Japanese newspaper with a headline calling Sun Yat-sen the leader of the Chinese Revolutionary Party, with *ko-ming* being used for "revolutionary." The Japanese use Chinese characters for their main points, and the headline was entirely Chinese. The Japanese had taken these Chinese characters into their political vocabulary, which got great new infusions with the Meiji restoration, and they became metaphorized to mean "revolution." After seeing this headline, Sun said, "Suddenly I realized that's what I was." Sun had not used *ko-ming* before because of its monarchical associations. He still took this word at a literal level, as part of the vocabulary of political discourse in a political system he wanted to abolish. Suddenly he had an epiphany in which he felt that this term had been metaphorized and no longer meant only "exchange of the mandate," but could also mean "revolution," as in the Japanese usage.

The Japanese came in contact with all sorts of Western things abruptly in the 1860's and '70's, and had to look for new vocabulary. Because the Japanese monarchical system was quite different from the Chinese, they had never used *ko-ming*; there was no idea of the exchange of the mandate in Japan because that would have been considered a loss of virtue. For the Japanese this concept was subversive because of their ideas on the eternal nature of the ruling lineage. In the nineteenth century, they had to have a word for revolution, and they used *ko-ming* which did have something to do with drastic political change. Sun then adopted the Japanese usage and fed it back into the Chinese.

One has a vivid picture here of a term losing its old literal currency and taking on a certain metaphorical currency, but one also sees the continuity. It is still China all right, but one also gets a sense of change—particularly when somebody says in somewhat sardonic terms that Chiang Kai-shek "lost the mandate" in 1949. Such a person is obviously speaking in metaphorical terms, but his remark has a historical resonance. The recognition of the new kinds of connotations and associations of terms shows you the continuity of the situation, but also reveals the change.

New Trends in History

POCOCK

What you have is the history of universes of discourse, which are capable of enlarging themselves and taking on new dimensions in highly bewildering ways. It occurs to me that we have got away (thank heaven) from questioning the relevance of the social sciences to history. I can think of various sciences that have something to say to us, but I cannot think of one that can make a major formal contribution to an analytic problem of this sort.

HOFFMANN

How do you determine, for instance, why a certain word appears at certain times or what the word was used for without making some implicit hypotheses about the way in which a social and political system may have influenced the choice or the meaning?

POCOCK

I did not mean to say that this method ceases to be analogous to those of a social science; I am trying to emphasize their similarities. As I look around at my intellectual friends, I do not notice a bunch of specialists and think, "I ought to go to school to learn what those boys do." I did once wonder if I should take a year out to retool my linguistics, but then decided that linguistics was on a level of abstraction that would not be useful in what I was doing.

HOFFMANN

Perhaps this is something on which you should start a battle with Frank Manuel. It seems to me that some of the work being done in France under the rubric of structuralism, which is perhaps not yet accepted as one of the social sciences, is trying to go in this direction, and the people doing this are not historians. It seems to me that it is very hard for a historian to get completely away from the rest of the universe.

POCOCK

We might talk about the *structures mentales* and how these resemble the language patterns.

HOFFMANN

Is there such a thing as intellectual history?

POCOCK

There is a set of techniques that keep me busy, and I cannot think of a better name for them than that.

HOFFMANN

Earlier we heard the lament that social history was a residual category that concealed a large number of totally different enterprises. The unity is just as artificial, it seems to me, if one puts together under the name of intellectual history your kind of work, Frank Manuel's, some of Erik Erikson's, and some of the more traditional work of literary history. People ask of a certain body of material totally different kinds of questions. The only unity is provided by the raw material, which is a very primitive sort of unity. On close examination, it goes off in all directions.

POCOCK

I agree. There are a lot of intellectual historians and there are historians of intellect. There does not have to be a unified discipline called intellectual history, because one uses the intellect to do quite a lot of different things. That is why I am unhappy about the *structures mentales*; they sound as if they are coming up with something like those unified world pictures which are supposed to underlie all the aspects of thinking in a particular society.

BEER

Mr. Pocock, could you tell us more about what your language system is? How do you know one when you see it? Is it a psychological entity or is it a set of logical implications?

POCOCK

To my way of thinking, it is an institution. I can recognize the specifically political vocabulary of a society, in some cases with more difficulty than in others. To begin with, it is a linguistic system, a bunch of words.

BEER

But what makes those words a system?

POCOCK

It is a system because the vocabulary is generally understood to be linked together in ways that are varyingly explicit. Using a particular language commits one to certain consequences. You tie yourself up in various ways; you set yourself going along various paths you did not know were there.

BEER

Because of logical implications or because of some other reasons?

POCOCK

One can certainly imagine that logical possibilities known to some political societies are not known to others. I am not sure how far one should press the word "logic." I am content to say that a language system is made up of related words.

HOFFMANN

Two questions on this. You say "understood," and I must counter by asking "understood by whom?" This question brings us back to the social sciences. What is the limit of this language system? Who are the participants, who are the people who understand it? The answers to these questions would, it seems to me, vary from place to place.

Finally, I would want to ask you, "language used for what?" Language has many different functions. Sometimes it is used for what I would call purely intellectual purposes—for understanding, for advancing research or thought. Other times it is used for political consequences, by politicians who want to get something out of using it. Doesn't one have to look at the functions that this language fills? And if you diversify functions, doesn't the notion of the system seem to disintegrate a little?

POCOCK

Answer to question one: That is a question about the political structure of the society. Answer to question two: Yes, of course there are two sides to this medal. One is communication within the political system, and you have to specify the political functions that words are being made to perform—they legitimize and they delegitimize; they orient and they authorize. At the same time, of course, you are not just looking at the consciousness of politics; you are also looking at the politics of consciousness. You do go off into the sort of area that Levenson studies—namely, the conceptual universe inhabited by these people, the consequences of their being the political animals they were. And at this point, clearly, you depart from the language system into the culture. But that is a thing that political animals visibly do. And all kinds of confusions follow from this point. On the other hand, I see the activity of talking about politics as developing into a number of different autonomous intellectual activities—aspects of philosophy, historiography, and so on. You can get to the point where there is an autonomous political science, but there is a truly autonomous political science only when its vocabulary can be isolated from the vocabulary of politics in general.

TILLY

I wonder whether you breathed your sigh of relief in escaping the social sciences too quickly.

POCOCK

I knew it was only a respite.

TILLY

I am puzzled by what you say about how you delineate the object of study. It seems to me that at two points you actually rely on the substitution of your own intuition for what social science can barely do. At one point, you identify something from political vocabulary and say that is self-evident. I am not quite so sure. At a second point, you base your judgment on the coherence or the interdependence of the vocabulary on some observation of how people employ the vocabulary. At those two points, it seems to me, you throw yourself right back into the central problems that political scientists and numerous other people have been contending with all along. In fact, you have somehow balled it all up together—a set of judgments about the characteristic political system that lies behind any set of utterances in an intuitive form. Do I misunderstand your procedure?

POCOCK

If balled up means rolled together, I am not sure I get your point. If you are asking me how I recognize any political vocabulary, the answer is that the vocabulary of politics, like the word itself, has a large range of meanings, and one gets to know them. How I knew it was political thought when I was reading translations by Arthur Waley is, of course, another matter. At this point, I simply applied some rough-and-ready concepts of what political ideas were —words about norms, about authority, about obedience, and about commands. I said "this is politics" and then made some more adventurous statements when I talked about Chinese political thought being concerned with how norms are transmitted and known in society. I seemed to be looking at a series of different images of how authority should be organized and, incidentally, different images of an authority structure as it existed and legitimated itself in time. If someone asks how I know these things are politics, however, I merely reply they are politics.

GILBERT

I would like to take up one question that was raised earlier. What is intellectual history? It seems to me that if we look at the traditional divisions of history—political, diplomatic, social, and so forth —the two fields that the nineteenth century left out were social history and intellectual history. We are meeting here to talk about new trends in history. It might be worthwhile, therefore, at least to keep in mind the question of whether these fields have merely changed their aspect in recent times or whether they are actually new trends, whether the intellectual history which we are cultivating is more tied up with problems of social history than the history of ideas of previous times.

I am interested in the question of the socialization of ideas. Historians seem to have entered a state that is rather different from the context of the nineteenth century, but this shift seems much more complicated to me than just an increase in concern for social history.

KRIEGER

I would argue that the new trends have been in the direction of trying to find some kind of process, some kind of relationship, between articulated ideas and something in the psyche of the society. This is implicit in Professor Pocock's attempt to get at the political language in which particular ideas adhere and explicit in Professor Manuel's analytical treatment of ideas by means of the personality. These are all ways in which the intellectual historian has defocused his attention from particular formal ideas, from the finished intellectual product, and sought to relate it to some kind of psychic or social or linguistic process.

MANUEL

I did not originally set out to do a particular kind of history of ideas. I knew that I wanted to deal with hard materials, with refractory materials, with difficult materials that did not obviously lead to a description of the feeling tone of an epoch. In my book *The Eighteenth Century Confronts the Gods,* I took the learned works of about two thousand eighteenth-century thinkers and tried to give meaning to and interpret the body of mythography they represented. Most of this literature is dull, learned. I found myself quite intentionally asking what I could make of it and how I could learn something new about the Enlightenment by approaching it from another vantage point. There was no unified way in which eighteenth-century man looked at myth, but I think I was able to express certain ranges of feeling tones of the age—attitudes toward love, toward reason, toward sex—by considering this body of material which would normally have been consigned to a history of classical philology. Important for me was the indirection of the method.

To some eighteenth-century thinkers myth was political history in disguise. I think I found that one of the eighteenth-century trends, a tendency toward the matter-of-fact, was manifest even in their mythography. I also began to find a psychologizing of myth in the eighteenth century.

I have been engaged on another long-term study, not wholly deliberately, and that revolves around the fantasy of utopia in Western thought, the perfect polity or the perfect state of being, the perfect state of feeling or the best commonwealth. This interest first started when I was working on St. Simon and I was intent on relating his utopia to specific social and political circumstances. When I studied a whole group of utopians in the *Prophets of Paris,* I found that there was a way of tracing relationships from one generation to the next. The use of the idea of utopia was important for me because it gave me access to a concept around which other ideas clustered. The intrusion of Thomas More's word "utopia" into the historical process was particularly interesting to me. Ultimately, I hope to do a history of the Western utopian imagination as one way of revealing changing feeling tones from one period to another.

Now you will ask me how the utopia is related to the society in which it was created. Every once in a while I hazard reflections about the relationships between the idea and other aspects of existence. I am fully aware of doing this. At other times I am content to map the territory, to plot the varieties of utopian experience in some kind of sequence. I feel that I cannot yet jump to an explanatory system, although I try to keep that objective in the background at all times.

A third type of intellectual history that is even more problematical interests me very much. I fell into the hands of a scientist —Newton. When I was working on the myth book, I got hold of Newton's theological, mythological, and chronological manuscripts. These works had previously been treated with disdain, and this refractory, ugly material appealed to me. I wanted to give it meaning in terms of contemporary chronological, theological, mythological interpretations. Newton happened to devote a good part of his life to this material. I began this purely as an exercise. But I found that Newton has a way of sucking people in. The problem soon became an attempt to make sense of this man who wrote so many words on alchemy, so many words on theology.

I went back to an essentially biographical approach very much under the influence of Erikson, but without accepting his categories or his language. In fact, my approach was sometimes much more early Freudian than Eriksonian. What it will give I do not know, but part of its wicked intent is to break down the idea of the autonomy of scientific invention; another part is to show that other elements of the human being were at play in making some of these apparently most abstract of all constructs. I hope that the ultimate product will be a presentation of a most complex mind-feeling activity. It has the advantage for me of giving me finiteness of dimension since the man is born and he dies, but it is a long life and a life during which there are many intellectual currents continually intruding to alter what appear to be some of his autonomous reflections.

Now this is heavily psychologized intellectual history, but it is no specific kind of psychology. The "gods" book clearly took its model from a Rorschach test. There is no question in my mind that I was using that model in a primitive way, but that was my purpose. The Newton study was a life history. I am now faced with all kinds of problems as to the relationship between the individual life history and the polity. I do make attempts to see how the organization of an intellectual system is accomplished in the social world. I do not think of these as abstract beings.

In a series of lectures at Stanford I examined philosophies of history and asked if there were some enduring patterns in the repetitive forms of historical philosophical myths. There I did not at any moment relate the inquiry to specific social or political reality. I do not at all feel, however, that a philosophy of history or a utopia is something that occurs in outer space. I consider mine a mapping operation, which is not to be denigrated at this stage in intellectual history.

Thus, different persons are doing different things in history, and the same individual is doing different things at different times.

BEER

How do you know when you have the feeling tone?

MANUEL

If someone took the two thousand treatises involving the nature of myth that were published between 1680 and 1800, it is quite possible he would come up with a configuration different from mine. Most people who have dipped into this field and read my pattern would say that it is at least plausible. For me, the great creative moment in doing this came when I made the groupings and felt satisfied that I saw the structure.

SCHORSKE

Was this configuration one of similarity in the material you were studying or was it associated also with a configuration in the social community that produced this material?

MANUEL

Both. There was really an echo. The real moment was when I heard this echo.

SCHORSKE

How big was the community? Was it the professional or educated community?

MANUEL

These books were read in quantity by about as many people as read Rousseau.

KAYSEN

Having got a reasonable pattern or feeling, are you interested in such questions as these: Is there in some sense a finite number of plausible patterns? Do those plausible patterns have relations to one another, or is a plausible pattern a purely personal aesthetic experience?

MANUEL

I would want to get away from the idea that this is solipsistic. It is conceivable, though dubious, that there could be more than five or six plausible patterns. Somebody might make four; somebody else seven. Other people have dealt with this kind of material and have come up with different patterns. I do not feel strongly about where they draw the lines. I do not think that one could take that position. There are two thousand of these treatises; each one is a complete thing in itself. My configuration-making need may be so strong that I cannot leave these two thousand dots hanging out there. The amount of similitude they have, in my contemporary sense of the similar, may reduce them to, let us say, five groups, and that is about where I feel satisfied aesthetically and intellectually.

LANDES

I am not sure that Frank Manuel is answering the question that Carl Kaysen asked, or at least not the one I would ask. You have reduced things to five patterns. Could ten other scholars each have had just as easily five patterns totally different from yours?

MANUEL

I could conceive of that perfectly well, although I do not think we *would* be that different. If somebody else went through this body of material at this moment in time, we would not be widely different.

HOFFMANN

Does this not depend on what kinds of questions you ask your material?

MANUEL

Certainly. If someone had read those treatises for what they reveal about economic history, he would have got a completely different pattern. I was obviously asking what they reveal about the feeling tone in terms of the way the age expresses its emotivity.

SCHORSKE

You have taken learned treatises and revealed their affective content. Do you then proceed to some other arena or is that the terminus of the inquiry?

MANUEL

That was the terminus for this inquiry, and I was perfectly satisfied with letting it end there, although this study obviously does not exhaust the possibilities.

KUHN

I am a little distressed, Frank, because I think you are giving more away than you need to. Let me just say that the method is probably stronger than your statement implies. In the first place, there is a long way between the two alternatives that **Carl Kaysen** posed for you—the one being that you felt you knew enough about what was going on to say that there are only five plausible patterns and the other being that you would say anybody working on these materials might get anything he wants.

From my own experience in dealing with other constellations of ideas, when somebody else starts working on the same problem, the differences between us are unlikely to be total. Rather, one of us may try to split one of the five categories into two and things of the sort. The historian's sense of the appropriateness of his categories is not purely subjective. When you pull a book off the shelf, you know which of the categories it is going to fit into and by god it does. You find the recurrent phrases that you have been looking for, and you begin to develop a reasonable amount of confidence that your configuration is going to hold up.

The other point I want to make is that when you learn to do this categorizing job right, you will not get totally different ways of sorting if you go in with a different set of questions. As Felix Gilbert suggested, people concerned with the history of ideas are increasingly asking questions like: Who are the social groups who hold these ideas? If, indeed, one has patterns of ideas that make sense, they probably refer to people who can be sorted out in social terms. If you then read the same materials for economic history, you are likely to find the same social groupings also reflected in the sort of economic historical doctrine that comes out of them.

I think it unreasonable to defend the categorizing method as a stopping point and to say that if you go in with different questions, you will get totally different groupings. I would expect to get correlations among the groupings of, say, the intellectual and the economic historians. These enterprises do and should interpenetrate and interilluminate one another by pointing back to the same clusters of individuals in which they are rooted.

POCOCK

The groups will overlap because the same individuals will be interested in different things. The historian now specifies the social group much more functionally than he once did—it's the cultural group, it's the political group, it's the scientific group, and so forth. We are emphatically less interested in questions about whether this is the rising gentry. For the most part we are looking at people playing language games, in which they are involved as social beings.

MANUEL

I do not call them games in the modern sense, because these are the real things to me. If one had the data, one could analyze who these writers were. This would be difficult to find out, but it would enable one to see if there was any correlation between the different theories that came from all over Europe and the writers' social status. I do not think the answers to this question would be particularly exciting.

SCHORSKE

If there is to be any social function for ideas, then it seems to me that the identification of the persons who are their creators and purveyors becomes quite essential. In the social history of ideas, you begin with the people who are making the ideas and then go to the community. You try to find its character so that you can perhaps get some new kind of reflection from that social community on the ideas.

GILBERT

I wondered whether Professor Schorske could give us some idea of what he is trying to do in studying Vienna at the end of the nineteenth century?

SCHORSKE

I am not awfully articulate about my own work, but I will try to explain it developmentally. In teaching intellectual history I always felt that I was getting into a hopeless morass when I reached the late-nineteenth century and that morass became my interest. After Nietzsche, all the categories for dealing with so-called high culture in Europe essentially break down. We no longer have those handy concepts such as rationalism or romanticism to serve as a convenient base from which to explore and illuminate concrete phenomena (even though the concepts in and of themselves obviously do not have totally binding validity).

What began as a defect in my teaching became my concern as a scholar. I first tried to find general concepts which would govern several fields—philosophy, literature, fine arts, and so on—but this approach proved to be futile. I then thought that it might be better to explore a single social unit out of which the new culture emerged. After exploring several cities in seminar—London, Paris, Berlin, and so on—I concluded that Vienna offered an extraordinary opportunity for research. It contained highly creative people in a variety of fields—psychology, economics, philosophy, music, architecture, literature—at the end of the nineteenth century. Thus, I was driven to Vienna not because I loved Vienna, but because it looked like a good unit for studying the break from nineteenth-century realistic thought and the emergence of new twentieth-century intellectual forms in one sphere after another.

I have wasted incredible amounts of time trying to bridge the divisions of activity by which these people identify themselves. I tried to descry intuitively commonalities in the structure of thought between, say, literature, philosophy, and economics, and I failed. I got a lot of nice hints on the common, characteristic concerns or attitudes among intellectuals in different spheres of thought and art, and this approach enriched my feeling for the period, but it did not reveal any recognizable *Gestalten* shared by these disciplines. Hence, I had to go back and explore each discipline in terms of its own inner logic.

While making this exploration, however, I also tried to use the biographical approach on a rather large scale—the methodological assumption being that any high-culture artifact, any product of thought, is a product of a life lived in a social matrix and that the biography is therefore the proper mode for connecting the creative person with the society out of which he emerges. I took single biographies of cultural innovators and tried to find out where they led me in terms of discovering what communities these people lived in. They lived in at least two: in a professional community, but also in a social community larger than the professional one and defined in a different way.

What was their social community? It was an upper-middle-class cultural subgroup in Vienna, of which I have developed a kind of multiple, many-faceted picture. This subgroup can be analyzed from the point of view of its political culture, its religious or parareligious culture, its family culture, its mode of entertainment, and so on. I have tried to sketch a kind of broad canvas to which the evolution of each of the specialized professional groups can be referred back for historical contextualization. I have watched the individuals arise out of this milieu—observing what they take with them out of the milieu as they begin to develop their own intellectual activity.

I have now developed an assumption—that the high cultural productivity was based on a revolution of falling expectations in the class which produced the culture. Most of the intellectuals began their secession from Austrian liberal culture and their creative substitutes for it not as intellectuals alienated *from* their class, but *with* it. The class to which they belonged was extruded from participation in political life, from the exercise of political functions that they regarded as quite central to their shaping of the historical future. The defeat of a certain set of historical expectations, therefore, provided the ground against which the new culture was devised; and the specific form of the new culture is derived from this fact.

Internalization is, in a certain sense, a key to this process, although it is not enough. Fluidity, a sense of flux and metamorphosis, is another key. A characteristic may not mean anything at all until you see concretely how it is manifested in the various disciplines. There is a lot of difference between fluidity in painting and architecture, and fluidity in music, and fluidity in a psychological model, but the characteristic does appear in various branches of thought and art.

The basic phenomenon is the secession of the intelligentsia from the thought-patterns that belonged to the main stream of the paternal culture; it is therefore also the revolt of one generation against its predecessor. Each professional community begins with a concern that is drawn from an arena larger than its specifically professional concern. Almost everyone who became culturally creative in this first generation—which included Freud, Otto Wagner, Camillo Sitte, Gustav Mahler—had been extreme German nationalists and Wagnerians during his university years. These men first opted out from liberal culture when they were forced by political circumstances to choose between being Austrian liberals and being German. This initial student-generation choice then became transmuted. Political consciousness gives way to theatrical consciousness, an aesthetic mode of community formation. This then passes on to the plastic arts, to city organization, and so forth, where theatrical models are again used. You can watch, in that sense, the large community as it defines a general intellectual milieu, out of which one or another substantial figure will emerge. I start with a network of considerable interdisciplinary complexity that creates a professional community, and out of that professional community I take one major figure who is a carrier or refractor of many of the things that are in the larger milieu.

POCOCK

The relevance of what Professor Schorske had been saying before lunch has become clear in listening to him just now. We were discussing earlier the question of how far it might be desirable to specify the community or group engaged in a particular intellectual activity. Professor Kuhn has a community of scientists talking in scientific language using scientific paradigms, and I am interested in political societies talking a specifically political language, and so on. Professor Schorske raises the possibility of the same group playing a number of games simultaneously. He has indicated, it seems to me, something quite useful to do with biography—tracing an individual's passage from one game-playing group to another.

Is it possible to retrace one's steps and look at philosophy highly refracted as it was by the physical scientists in the Vienna circle?

SCHORSKE

Yes, one certainly can do that.

KAYSEN

How would what you do with, say, the literary groups differ both in methodology and intention from what a literary historian does?

SCHORSKE

This is a good question, and it could apply equally to every group I am studying. There are literary critics who do things with much the same purpose in mind that I have. They happen to be living in literature departments or to wear the label of literary historian. I have been learning a great deal from the sociological literary historian, and he, too, seems to find my work congenial. Thus, I would say there is no difference between me and a literary historian when the literature selected for treatment by either of us has a high utility in relation to the other elements in the cultural configuration that is being explored historically.

In a way this points us back to Tom Kuhn's memorandum where he speaks of the square pieces of the puzzle that have different colors. A historian has a certain purpose and so he picks out things that are germane to that purpose. I am trying to illuminate the function of literary culture in this society and the way in which the society affected the literary culture. A literary critic could, however, perform several alternative operations that would not be of interest to me. He might, for example, analyze a poem structurally for purposes entirely unrelated to the intellectual content of the poem. I am interested in the values in the poem as they reveal something about social change and development. Formal analysis of a poem comes into my work only when it is related to making more explicit the historical content of the poem. Thus, the formal analytic skill I would employ or command is relativized to the historical purpose of illuminating the cultural or social content.

BEER

How do you determine function?

SCHORSKE

I will give an example that I have worked on recently—the architecture of the first modern art exhibition hall. This hall was designed by a new architectural movement. It called itself the Secessionist Movement, the Austrian equivalent of Art Nouveau.

The group that built the exhibition hall began with an ideology about the function of art. This ideology was formed by artists and non-artists—ex-politicians, literary people, lawyers, and others. I am relying here on conventional verbal sources—statements of patrons and supporters or simple propagandists who articulated their aims and claims about the function of art.

I then look at the design of the building, at the way in which the space is organized, and try to determine the extent to which the organization of the space resembles what has previously been museum space or exhibition space. This particular space is unique in the architectural history of Vienna; there have been precedents for its fluidity, but not in museums.

This uniqueness in architectural form can then be tied back to the ideology of the new movement, with its notion that it was not going to take the old art any more. Its slogan was: "To each time its own art and to art its freedom." Freedom implies fluidity, exfoliation, possibility. Nobody knew what art was going to look like in the future; consequently, the architecture to house it had to be open-ended.

The literary sources also tell us that the museum is a refuge, a temple; that art would provide not an articulation or reflection of modern life, but a sanctuary from modern life. In art a man can escape the hurly-burly and the traumatic experience of being slapped around by the conditions of contemporary existence; he can find surcease, repose, composure, and the like. Thus art becomes a surrogate religion and provides a certain set of psychological functions that are quite different from self-representation. These functions are reflected in the conception of the secession's exhibition hall. The basic form of the museum is, therefore, a temple; it is modeled on a certain combination of Egyptian tombs and a Greek temple in Sicily. The museum expresses the ambivalence of the group that built it—on the one hand, affirming the openendedness of modern man; on the other hand, providing a refuge from its consequences.

You can analyze a building, a painting, a statue, to see what it contains that enlarges iconographically the discursive knowledge that one can acquire from literary sources.

GRAUBARD

What is it about this kind of research that makes it research that we would not have seen done fifty years ago? What you are doing does remind me of a certain kind of nineteenth-century research; and here I think of a man like Burckhardt. What is missing in your account is a political component. The word "politics" did not figure in your initial remarks and nothing you have said since has made it figure.

SCHORSKE

It did, but only in passing. There are, however, several places where it does figure. Politics has not been dropped as a consequence of an interest in social and intellectual history. On the contrary, something like the Greek concept of *polis* has been restored, and politics has now become a far more holistic and complex subject than it was earlier thought to be. We have lost the legal, constitutional, institutional, and even political party focus by broadening it. The *polis* is seen as a cultural entity, as a whole.

GRAUBARD

Nevertheless, the questions you have asked have to do with the kinds of things that have not ultimately seemed so important in a political age; they have to do with the relations of an individual to his class position and so forth.

In the nineteenth century, intellectual history was not simply something left over; it was much more closely allied with political history. Burckhardt was a great intellectual and cultural historian of the nineteenth century, and it is not entirely an accident that he wrote both political history and cultural history.

You seem much more interested in intellectual-social history. This shift would seem to me to reflect the kind of change that we were talking about at an earlier stage in our discussion. One cannot get around it by simply saying that the politics of the nineteenth century is now the society of the twentieth century. The questions Burckhardt asked had to do with political style in a much more explicit way than do those you are asking.

KRIEGER

That is simply not my view of Carl Schorske's conception. The function of politics is very different in what he is doing from its function in the intellectual history of the nineteenth century. Politics becomes, in Carl's view, one of the key strands in the decline of expectations; politics becomes part of the lens through which his groups view their society and their culture. Politics becomes not a segregated part of the system, but part of the prism through which this group sees its whole life. Thus, the function of politics is different, but still central.

ROTHMAN

This art museum analysis intrigued me because there was a convergence of intellectual history and social history in a suggestive and methodologically somewhat unique way. You are all involved in organization and allotment of space in an art museum. I often find myself highly involved in questions of organization and allotment of space in very different kinds of institutions—prisons, almshouses, or mental asylums—where people are forced to create an institution to service a social system. Basic decisions about who will sleep where, how the inmates organize themselves, how they work together—these become critical in using institutions to illuminate the larger system. If we grow more and more sensitive to these vital lines of evidence—something which Burckhardt was not, I think, doing in quite this way—we will manage to make intellectual history something that tells us an enormous amount about a society as well as about the unique participants.

SCHORSKE

Burckhardt uses a cross-sectional technique. In every chapter of *The Renaissance* there are about one hundred words, usually very close to the beginning, which tell the reader the way it was in the Middle Ages, and another hundred words that tell him what the modern world is like, and sometimes another hundred that relate the two. Between the static backdrop of the Middle Ages and the scrim of modernity, another aspect of the Renaissance is displayed as a *tableau vivante*.

TILLY

I should not think that Carl Schorske's procedure would be so good for giving accounts of change except under one special circumstance—where there is a sharp generational shift of some sort and there are cohorts whose experiences can be treated as homogeneous and persistent in their effects. Many changes we deal with in social life are incremental; there is no change in any particular person's life which is of great importance and yet collectively there is great change. Shifts in vocabularly, shifts in political notions, sometimes occur this way, as do changes in tastes, language, and demographic structure. It would be difficult to apply this model of well-defined, homogeneous experience to incremental change.

SCHORSKE

The generational effect is definitely part of my postulate. Moreover, I have two generations: one generation that made the break with their fathers' culture and a second generation that grew up with the achievements of the first as a base. The second generation did not face a big social crisis; that had been taken care of by the previous generation. The second generation was far more rationalistic, yet more explosive, than the first. From my point of view, the second generation would be unthinkable without the liberating work of the first, and the first group could not conceivably have moved to the second stage because they all had terrible crises of guilt in facing the consequences of what they had discovered.

As a matter of fact, for me, the Oedipus myth in its entirety became a model for this earlier generation—not just the part about killing your father and sleeping with your mother, but Oedipus as King of Thebes, a political figure, who ends up the wise man, the blind seer, and serves then to reconcile others to fate, to the laws of destiny. Without exception, the personal experience of all my major people in the first generation was very similar to that of Oedipus, when the myth is taken in a quite complete sense. There is a father-slaying, a tremendous sense of breaking open new vistas as a consequence of this, and then the realization that these new vistas have opened up an awful Pandora's box.

Freud's personal experience of suddenly confronting the full consequences of having engineered a generational revolt is just one among many. At a certain point freedom turns on its own tail and becomes terrifying to people. They had no idea, for example, of how painful the discovery of sex was going to be.

GILBERT

We have a picture of certain aspects of our intellectual history that are connected. I would now like to go on to some of the other memoranda.

MANUEL

I would like to hear from Mr. Herlihy as to precisely what he would put onto a computer card for a particular medieval charter.

HERLIHY

By using the computer, medievalists would be able to utilize data collected by other medievalists far more efficiently than they now can. At present, people go through voluminous quantities of documents, their work largely appears in tables that are much reduced, and they keep the rest of this great mass of material in their filing cabinets where it never serves anybody.

The computer might permit historians to work more as a group. Obviously what is recorded has to be as broad as possible, and obviously there are technical and financial limitations here. We are looking forward to a kind of open-ended machine-readable edition.

Charters have the great advantage of being homogeneous, as there are a limited number of types. They are legal documents and therefore have a legal structure. Despite the vast number of centuries that they deal with, one can quite easily relate a sale that is dated 750 with one dated 1150. They deal with the same sort of buyer, a piece of land, and a seller. The idea would, therefore, be an open-ended edition that would be partial in the beginning, but would allow additions of further information from the same charters at a later time.

In other words, initially we would have a register of the individual charters giving essential information. The great difficulty is proper names and place names. Many of the place names are not even known, and this obviously presents problems. But if we were to leave out all the place names, it would be possible to have a description of the charter plus an identification number. Other historians who wanted particularly to study place names could then utilize that same identification number and add further to the material.

The coding system must be open-ended and empirical so far as this is possible. It should reflect the document rather than the analytical interests of the person who does the coding.

I think it is fair enough to consider this problem as being analogous to the problem of defining what a critical edition was, which obviously had to be settled in the classical age of historiography: What should a critical edition carry? Should it show mistakes? In utilizing these machines, we are doing the same thing; we have to determine the essential characteristics of a critical edition that would be machine readable. Initially all the data cannot be recorded, so the first task is to inventory them. Further additions both in terms of material carried and also in terms of additional charters can be added later.

KAYSEN

How many words are there usually in a charter?

HERLIHY

It can vary a great deal, but I would suppose there are several hundred.

KAYSEN

Then the obvious thing you want to do is to put the complete charters on the machine. If you had ten or twelve thousand words per charter, it would get to be hard; but this is perfectly feasible. With the complete recording, you could fool around with the relations between different coincidences in the texts. Before you get very far with this project, you should think seriously not about coding the charters, but about reproducing them.

HERLIHY

I am technically ignorant as to the exact possibilities; I do not see how it could be done.

KAYSEN

It is very simple. Under your scheme you have an entry which is a number and you associate that number with something—a library, a place name, whatever. Then opposite that entry you have a code which tells you who the buyer was, who the seller was, where the sale took place, and so on. Let's suppose that code had fifty entries. Why not think big and have a thousand entries instead of fifty—those thousand entries being all the words of the charter.

HERLIHY

Wouldn't that require a great deal more work from the person who is doing the coding?

KAYSEN

No. You can photograph the charter, and a machine can translate the photograph into a code. If you are talking about a factor of 100, never throw anything away. A factor of 100,000 might be expensive or impractical, but a factor of 100 is neither.

BENSON

When the whole text is there, you can use your own codes recognizing full well that ten years from now a new concept will come about that does not fit into the existing code and that you will be able to add it to the original.

KAYSEN

The whole data-bank business consists essentially in not throwing away any information. Until now, it has been technologically and economically impracticable to save all the information of, say, the censuses. These have been tabulated and then the original information has been thrown away. If you look at the 1927 census, there are a lot of things that you could have learned from the original information that you can no longer learn from the tabulations. It is no longer necessary to lose this information.

BENSON

Changes in technology are making the old key punching obsolete. The key-punching costs, the critical item in the entire operation, are being reduced. The cost of recording election statistics has been reduced at Ann Arbor by about four or five times in the last year or two.

Without, I hope, sounding too much like a technological determinist, I think that the nature of historical research is going to be transformed in large measure because of the technological possibilities that now exist. It is no longer necessary for historians to examine all the documents physically. Thus, the main function of the historian will not continue to be one of data processing, which seems to have been the historian's main function in the past. I would guess that historians have tended to spend 90 per cent of their time in primitive and inefficient data processing.

ROTHMAN

If the information is available on tape, how will it come out?

BENSON

To my knowledge, the simplest form now is magnetic tape from which the computer will print out the whole text.

HERLIHY

If one wanted, for example, to study the word "fate" in several medieval documents, could one ask the machine to print out the pertinent paragraphs?

BENSON

That poses no technical problem at all.

KAYSEN

For a fairly rigid document the machine is an efficient form of scanner, but the human being is much more efficient than a machine for reading. For example, the technique called content analysis is terribly inefficient. Any reasonably intelligent person who knows "the field" can do content analysis with a high degree of accuracy much more quickly by reading than by going through the process of content analysis.

BENSON

That is an old-fashioned view of content analysis. Some months ago a symposium at the University of Pennsylvania demonstrated that vast changes have taken place in content analysis. At Ann Arbor, there is an interuniversity consortium for political research that is creating a data archive of political statistics, broadly defined to include demographic data.

In principle, the same thing can be done for historical data.

Millions and millions of words are in machine-readable form, but are not now available at a central place. One feasible objective is to create a data bank of texts. Manuel's two thousand utopias, Levenson's Chinese poems, and a whole set of other things would be available at this central bank. All those notes stored away in people's filing cabinets can be centralized. This would give researchers access to incredible amounts of material that could be used without going through every single step in the process from raw material to getting access to it.

The American Historical Association has a subcommittee on collective biography. It hopes to create an archive that will contain in machine-readable form all collective biographies of whatever kind. These would be available to anyone for the cost of the tape or reproduction.

KAYSEN

I am going to persist in being old-fashioned. It is going to be a good long time, perhaps twenty years, before Frank Manuel's two thousand texts can be read by a machine to any purpose.

BENSON

It depends on what you mean by read. If you are interested in a subset of the two thousand utopias, a machine would enable you to locate those of the two thousand that are of interest to you.

KAYSEN

That is the part that I would want to deny. Work by juristic programmers and machine-language translators shows that deeply serious problems arise as to whether you can get a convenient index once the document does not have a rather rigid form. If it has a rather rigid form, then one can do all the things Lee Benson has been suggesting.

BENSON

Frederick Jackson Turner's famous essay on the significance of the frontier in American history has been abstracted by machine to one page. Most of the experts on that subject agree that this page is a good summary. Obviously there are sophisticated relations in the essay that the machine did not get, but that is merely a question of what you want to use the machine for. The machine frees the historian from collecting his own statistics. A machine can, for example, scan thirty-six different manuscript collections and tell me every letter in which John Quincy Adams is mentioned in connection with the slavery issue.

STONE

But that is an index.

BENSON

Yes; I am now only talking about indexing or abstracting.

POCOCK

I thought you were talking about much more than that.

BENSON

A man at Yale has done a "computerized" content analysis of Republican Party and Democratic Party platforms from 1848 to the present, using some complex concepts. In some respects I could get more and he could get more than the machine could if we read all the platforms and coded them laboriously. But the work is so dull and deadening that you couldn't find a cretin who would do it by hand.

It is true that machine translation was one of those dead ends that everybody rushed into. On the other hand, many people have been experimenting and doing interesting work with content analysis of complex literary texts and not just simple crude frequency counts. As historians, we should be alive to this field because we always engage in analyses of texts.

POCOCK

I would like to take a Luddite position here. I am rather unsure about what all this machine technology does to the intellectual and psychological processes by which I formulate questions. I am prepared to say that in a lot of fields once a good question has come into your mind, it does not matter if you have a reasonably finite amount of data to tie it up with. The Luddite in me gets frightened about what this immensely complicated process will do to my question-formulating powers. I can argue, I suppose, that I am a Gutenberg man. The things I like are in books; I shuffle through books and find something that I had not thought of before and I go ahead.

The question that comes into my mind may well be one that I could easily apply to getting vast amounts of information out of a complex indexing system. If so, all right. But it sounds to me as if any question that I ask of a complex process like this would in some way or other have to be filtered through all the questions in all the minds of all the people that made all the indexes, punched all the cards, programmed the machine, and so forth. If they and I were cooperatively engaged in the same highly conventionalized, highly technical enterprise, that situation might be fine. But it is perfectly possible in the field of intellectual history that there are a lot of books, a relatively small number of intelligent people, and a lot of fools. And I certainly do not want the fools to index the data for me, no matter how carefully and thoughtfully they have done it, because I should have to spend too much time feeding my brain through their machine and waiting for it to come back to me.

KUHN

By putting the entire text in, you are not the prisoner of any fool.

KAYSEN

It is a little more complicated than that, and I think there is something in Mr. Pocock's Ludditism. We have all presumably mastered an old technology—reading a card catalogue. Although I can read a card catalogue, I still find it useful to walk in the stacks. One learns things that way which you simply do not learn through a card catalogue. The machine, as we use it now, has some of the characteristics of a card catalogue. There are ways of using the machine that would be like walking through the stacks, but this is difficult. Our children will not have this problem.

KUHN

We do not know much about how creative imagination works, but I know my ideas come from turning books over and just browsing. Unexpected things crop up this way that I would never have got otherwise. Machines can index, and in the future historians won't waste so much time browsing over certain things, but the idea that machines can do this browsing process for us is, in my opinion, totally false. I think I am on Carl Kaysen's side.

BENSON

I have to agree with you. I did not mean that machines would become a substitute for browsing. I just meant to point out the utility of the machine as a locating device.

TILLY

In a modest way, I have been trying to solve some of these problems, and I cannot agree that the main utility of the computer is as an indexer. I want to identify one other kind of problem in which machines make an important difference.

Almost all of us—certainly all of us who come to classifying or counting relatively early in our analysis—do some reduction of data. The typical form of this in my profession is coding of one kind or another. Our usual vision of the computer stealing away our data is such that we code in the old-fashioned conventional sense—that is, we reduce occupations to nine categories and residences to twelve, and then we can no longer retrieve the information nor find distinctions. But the whole advantage of the shift from the crude machines that I did my graduate work on to the computer is that it is now possible to put the qualifications in the record so that one can go back and second-guess the initial reduction of data. In my own work, I find that this is the only way to proceed.

I happen to be coding descriptions of what I call political disturbances, which are extremely risky. The real importance of the computer here is not its capacity to index, but its ability to recall and then recode. We are using a program that permits us to do reliability checks on our coding. We can see how much coders agree on certain judgments and at the same time call back the prior set of comments that the coders made at the time that they made those judgments. Thus, this gigantic machine gives us a way of second- or third-guessing on sophisticated classifications.

KUHN

Interesting work is going on as a byproduct of this technological expansion. People are beginning to recognize how thinking about using machines to solve a problem changes the conceptual structure with which a person works, even in the absence of a machine. The work I am doing at the moment is of this sort. It is model-building; I am trying to play a game on a machine in order to discover enough about the game so that without using a machine I can begin to play it myself with texts. In a sense, a machine program can be a source of new ideas and provide new forms of serendipity, but I think we should go slow on the notion that large texts can yet be indexed successfully.

LANDES

I would like to ask Charles Tilly specifically what questions he is asking and whether the availability of the new data processing alters the precision with which he pursues those questions. What does his relationship with the machine do to his actual historical work?

TILLY

I have become interested in the changes in the character of collective violence and group conflict that occur during the process of urbanization and industrialization. I am working particularly with France and trying to look more generally at Western Europe after the French Revolution. I am using the machine to do an analysis of changes in the character of political disturbances involving collective violence over the period 1830 through 1960.

In a sense, it is the classic problem and certainly a very old problem: How has the character of French politics changed? How do we account for the volatility of the French? Who are the French rebels? I have really come at this as a sociologist who has learned to think about such problems in terms of what the sociologists call the theory of industrialization or development or social change, and that makes some difference in the way I set up the problem.

Given this broadly brushed background, however, I would like to say that the machines have made it possible for me to examine in a relatively uniform fashion about five thousand disturbances and to describe them in a relatively comparable fashion. This is not to say that the events themselves are similar, but that there are uniform descriptions of some thousands of events over a long span of time.

I cannot imagine how one could make the kinds of comparisons that I want to make without this large and stupid servant. Once I had decided that I had to look not only at large disturbances but also at very small ones in order to get the whole range, I was involved in such a complex data-collection and data-processing job that no group—let alone individual scholars—would have had the memory or the energy to carry it out. It almost had to be a machine job.

MANUEL

Can any of you conceive of using this machinery for putting documents into a chronological sequence when there are only a few benchmarks in terms of dates?

BENSON

My understanding is that pattern-recognition problems are being investigated, but that nothing of the sort you have in mind can now be done.

MANUEL

In the pretechnological age I spent six months on one incident in Tilly's sample of five thousand. I wrote an old-fashioned story about this one disturbance. What relationship is there between my spending six months on one story and Tilly's work on the entire sample? How does one relate these two experiences?

TILLY

I have read Frank Manuel's story and, as a matter of fact, there is a photocopy of it in my files tagged with that particular incident. In the short run, I think the material we assemble for the period 1830 to 1840 will provide a context of much greater breadth than he was able to achieve in his anti-machine story of the 1830's. At the very minimum, we will have some means of deciding on the representativeness of that particular event. I won't know until I start looking again at the conclusions he drew concerning the spirit of the time whether his conclusions about the general tendency in France from 1830 to 1850 correspond with my computer findings.

SCHORSKE

What else besides Frank Manuel's book is in this file?

TILLY

There are three classes of materials. There are materials from French archives; there are extensive contemporary press clippings; and there are the works of historians subsequent to the events. These are the raw materials from which we work when describing the political disturbances. In the background, there is another set of materials composed of demographic, economic, and industrial statistics of various kinds. These describe areas, communes, and *départements* of France, drawing heavily on French government publications. All these are in the files.

The machine record is a man-made digest of this material. It includes substantial transcriptions from the text at some points, but it does not include the full text of the documents, the newspaper clippings, or the secondary works we have consulted. Thus, I have not done the job that Lee Benson says is going to be done in the next generation.

At points where I thought we were likely to have trouble, where there was ambiguity, or where the coder himself was having difficulty, I provided a means of feeding in the text of the document he was interpreting and a description of his own behavior in coding the material.

ROTHMAN

I have not read Frank Manuel's story. Did you look at the social origins of the Luddites and then compare these with the social origins of the wider group?

MANUEL

No. It was purely a dramatic story about the breaking of one machine.

ROTHMAN

You were not interested in the question of who the Luddites were?

MANUEL

Of course. But Tilly is looking at the same kinds of phenomena, and yet I wonder if there might not be difficulty in adding together his presentation and mine.

I also wonder whether we will lose the emotive qualities of these confrontations if we stop doing this kind of *petite histoire* and resort primarily to quantification. It will still take about six months to write one of these stories.

LEVENSON

I would be interested in knowing if Tilly came up with any relative weighting that says that anti-machine sentiments are a relatively small source of grievance as compared with tax or religious grievances.

TILLY

As I have worked on this general problem, I have found one aspect increasingly interesting—and that is the actual form that collective violence takes and how it changes over time. The most frequent events that I have encountered in trends of the 1830's we could loosely call anti-tax rebellions and food riots. Both these forms of collective violence persist until the middle of the century. They break out in one last burst around 1848 and then very rapidly disappear.

As the century goes on, the frequent forms of collective violence become confrontations of formally organized groups, most having some kind of political identity. The typical thing is either a strike that turns violent or a demonstration that leads to some clash between troops and demonstrators or between demonstrators and counter-demonstrators.

This is not great news to students of French history, but the thing that impresses me is that it happened quite rapidly, and somewhat more rapidly in the cities than elsewhere. I am trying to correlate this change with changes in the country's political and economic structure. I suspect that if I could clarify this relationship through some comparison of who, when, where, how, and what form, I would be able to understand somewhat better the transition France was undergoing through the middle decades of the nineteenth century.

LEVENSON

Is there material for your "why" questions in your machine, or do you include material just for the "how"?

TILLY

The material for the "why" questions is in the machine. I have at least a way of examining some of the obvious things—like how fast the industrial labor force formed, the extent to which workers in one area rather than another concentrated in large factories, how fast people were leaving the *département,* or how fast people were piling up in the cities.

There is a serious problem in that the "whys" may turn out to be things I have not indexed at all. When I started this rather large process of collecting data, all I could do was to inventory for myself the kinds of information that were likely to be both accessible and relevant to most of the alternative notions about how political conflict changes over the course of large-scale structural change.

HOFFMANN

Do you also have for each one of the hundreds of instances of violence that you deal with the data on who the exerters of violence were?

TILLY

Not for all, certainly.

New Trends in History

HOFFMANN

I was interested in what you said about anti-tax riots, having once written a book on the Poujadiste movement, which was not taking place in the first half of the nineteenth century. You have cycles here, and it would be interesting to know with some precision who the anti-tax rioters were in the early period. I have a sneaking suspicion that the workers did not engage in anti-tax riots simply because there were so many grievances, but because somehow taxation was not a major one. More critical were problems of wages and political power.

TILLY

I do not think that is the case. The taxes against which people rebelled most actively were, in fact, consumption taxes of various kinds. The evidence I have so far is that the artisans of the time were extremely active in anti-tax matters. I concede your point that there was a recrudescence in the 1850's.

HOFFMANN

Do you have comparable information for all the riots?

TILLY

No. I have a characterization heavily biased toward government observers, newspaper observers, and the participants in the disturbances. In a good minority of those for which there is archival information, I have more detailed enumerations of those who were apprehended or identified. That certainly is a weak part of the analysis. I have fragmentary although not negligible information on the participants.

HOFFMANN

To some extent this is an answer to Frank Manuel's earlier question. You will always find room for both kinds of work: for detailed studies of one particular incident and for correlations among enormous masses of data.

GILBERT

It might be most interesting and valuable to apply modern methods, particularly the computer, to an analysis of last wills in the period of the ending Middle Ages and the beginning of the Renaissance.

If such an analysis could be extended over a longer period we might get data which show something about the development in the strength of family feeling and perhaps also the development in the strength of religious feelings and religious attitudes. But the last wills, in Venice for instance, are dispersed over a large number of very different files, and some particularly lengthy and rather technical investigation would be necessary before all the available documents are assembled. I mention this to illustrate that probably the kind of investigation which can be undertaken in early periods is very different from that in more modern periods where the files are much more uniform.

SCHORSKE

Has any new form of presentation emerged from the new kind of work you are doing? I can imagine years of research taking the form of a single page. Is it possible that there will be a new economy of presentation?

TILLY

Of course. It is, for example, quite likely that the results of some of these inquiries will be negative. Not only to save the investigator's self-esteem, but also to warn off other investigators, it might turn out to be worthwhile reporting negative results.

Thus, I would expect one new output of this kind of inquiry to be some larger reporting of negative results. Second, I would expect methodological discussions to become a more substantial part of the historical literature than they have been so far. This might be outside the journal as we understand it, like the semi-formal publications that come across one's desk. Third, I would think that a series of tools will begin to be published—code books, guides, and so forth. The Henry-Fleury manual for the study of parish registers is only one example. Finally, I think there will be increasing pressure to make the raw materials available in a form that is accessible to other scholars.

Thus, there will probably be strong pressure among historians to coordinate their labeling and to pool their data. So I expect something like tapes or IBM cards or printouts to become a more common form of historical publication, although this will not be publication of a kind we are now familiar with.

BEER

At this point, I see something that is fairly new in the trends in history; otherwise I have not been impressed by the testimony as to great novelty. Intensive case studies will go on, as will comparative studies; but this quantifying thing is new. Much of what we have said echoes Seeley's inaugural lecture of one hundred years ago. The social sciences are providing us with some new ideas, but the principle of applying these ideas to the interpretation of historical fact is not all that new. I am not myself particularly convinced that we have come across anything radically novel.

POCOCK

I am not yet clear about how these techniques make us look at new data or make us ask new questions. To a marginal extent, obviously they do, but they do not seem to have drawn our attention specifically to new kinds of questions.

GRAUBARD

To answer Sam Beer, perhaps your position on our discussion is correct precisely because we have not yet come to one of the items on our agenda: namely, that what is new in history is the emphasis on social history.

LANDES

Is that really new? There was certainly an emphasis on social history during the 1920's.

GRAUBARD

No, there is a difference. Social history then was what was left over, and it was essentially descriptive. My vision of social history as I read the books of the 1920's is an account of spreading technology, the bathtub as it makes its way across the Rockies.

Its treatment of institutions was similar. In the last year and a half I have been reading about universities, both in this country and abroad. This is a vastly underdeveloped field. There are not even any good biographies on the so-called entrepreneurs who made America's universities. Moreover, all the questions that you would have expected to have been asked have not been asked at all. I would predict that a vast amount of new work is going to be done in this whole area of education.

We are also still fumbling around in our notion of the history of the city. Again there is something of a tradition here; you have an occasional good historical monograph. Nevertheless, the city has simply not been studied as an entity. I do, however, see the beginnings of a tendency in that direction.

The family, too, is in for totally new treatment. In my view, this is going to be at once a social and an intellectual study. Whether one likes the Ariés book or not, I think we all feel an enormous indebtedness to it for its attempt to treat the family in something other than a simplistic fashion.

One can think of a fourth category—the factory, treated not in terms of factory legislation, but in terms of the structure of work.

LANDES

But these new trends are just subjects.

GRAUBARD

They are not just subjects. Until recently our notion was that only certain kinds of organizations could properly be described as institutions and studied as such. This is why the movement away from politics is so significant. I would argue that education as an institution is no less deserving of study than any of the political institutions that we normally attend to. Again, I do not think education or the city was conceived in quite this way by a significant number of competent historians in an earlier period of history. Historians are suddenly compelled to analyze these matters, and they will have to depend, in part, on the methods of other disciplines.

The kind of intellectual biography that Frank Manuel is writing was not undertaken fifteen years ago. He has not seen important materials on Newton that were not available to scholars in the 1930's, but they were not asking the same questions of the material that he is.

BEER

I am not dogmatic, but it would seem to me that new trends in history come from a powerful new idea. Economic interpretation of history was such an idea. Perhaps psychology and social psychology will provide us with powerful new ideas if we can ever work them out.

HOFFMANN

It seems to me that these are new trends even in Sam Beer's own terms, because they represent an extension in the scope of the questions which social scientists and historians are now asking.

Incidentally, on this rather futile debate about whether or not we have moved away from politics, I should think we have moved both away and not away. One simply has a different conception of what politics is. Carl Schorske was saying earlier that it is a conception in which one begins to understand that the educational system of a country is as much part of its political system as the work of the legislature.

Stephen Graubard listed mainly new schools of social history. One could add, as he did incidentally, the kinds of questions raised by psychology. Here one brings up the possibility of a renewal of biography. The questions to which the social sciences are drawing attention are almost all, by definition, present-oriented. Today we know that the educational system or the city is important. When one projects this backward, one must be careful not to read into the past things that are important today, but may not have been then. In this respect, I should think historians could play an important role by moderating some of the influences coming from the social sciences. Social scientists have a certain—almost inevitable—tendency to carry anachronisms with them when they go to work on history. So it seems to me that this is a fruitful area for collaboration. Historians are discovering a new area, and the discovery comes because other people have developed concepts or questions that the historians sometimes have not had the imagination to think of themselves. On the other hand, historians have a sense of the past as past, which social scientists only rarely have, either because they do not have the tools or because (especially in this country) they do not realize that historical materials cannot be used as if they were discrete.

BENSON

You do not want historians to do theory?

HOFFMANN

Why not?

BENSON

I asked that question because the role in which you are casting historians seems to be particularistic—one of clearing up anachronisms.

HOFFMANN

No, not at all.

BENSON

I find it difficult to believe that any significant new developments are possible in history if the organizational structure remains more or less the same. Some new organization has to occur; new functions and new roles must be developed. Substantive studies are essentially trivial if they are done simply for the sake of identifying the men who held power in the United States in 1835.

HOFFMANN

I could not agree more; but then this has always been true.

BENSON

Is there something in the nature of studying the past that prevents one from conducting this study in the same way that one would study contemporary events? That seems to me to be the unresolved question which we have been arguing about.

HOFFMANN

As a nonhistorian, let me ask you a question. Don't you have a feeling that what survives of the enormous amount of material produced by historians is precisely that work which tries to ask the kinds of questions you are interested in? Is it not conceptual history that survives?

ROTHMAN

I should like to return to some points in Stephen Graubard's remarks, which I think have been misunderstood. It is not that the school was never studied before; it is not that the university was never studied before; it is not even that the family was never studied before. But when they were studied, they were studied from a peculiar, unique, limited perspective. The host of volumes that have appeared on American schools is incredible. Yet they were all done from what could, perhaps, be labeled as an administrative perspective. This perspective would no longer hold. It is not so much that we have changed the cast of characters, but that we have begun to ask different sorts of questions.

GRAUBARD

I have been reading mostly about universities. In this literature, the university seems to be a rich old lady who has asked a group of admirers to write for a certain sum or for a ceremonial occasion the history of the institution. There are exceptions, of course, but they are few. I cannot, for example, think of one important history of Columbia University.

SCHORSKE

My comment concerns a remark Stanley Hoffmann made about our making theoretical contributions. I do not think of historians as doing this. History has always seemed to me to be a parasitic discipline. History borrows concepts from other fields; it does not really develop them.

Historians have a relation to their own classics that is to me very different from that of the concept-oriented person. When a field rests on the validation of concepts as its central enterprise, it tends to eliminate as no longer relevant to its on-going concerns classics that have been outdated even though they were recognized as great conceptual models in their own day and for many generations. Historians, however, still take up Herodotus and read him regardless of the fact that he wrote according to a set of principles that we do not believe in and quite missed the boat factually. Historians today would consider his principles of explanation wrong and his facts wrong, but when they close the book they say that it is a historical masterpiece.

BENSON

Why do they say it is a masterpiece?

SCHORSKE

That is the issue I would like to discuss. The appeal of a work of history has, in a certain sense, to do with its grammar—the way in which it gives plausibility to the empirical materials and the tightness of its articulation. I do not think this is artistic, although it may have some artistic ingredients. I should think it is more on the analogy of the structure of the language. To me, historians make meanings out of facts in the same way that we make sentences of words: through establishing syntactical relationships among the elements. This is why it is possible still to feel the impact of a book's mental structure, to be convinced of Herodotus' *Histories* when all its elements have become anachronistic.

GILBERT

I should like to pick up a few points from yesterday's discussion that seemed to me particularly important. I think it is something different to consider the nation-state as a temporary phenomenon existing during a number of centuries or to consider the nation as the one final element of history toward which everything else has been developing. If we talk about decline of the concept of nation, of national history, I think we are addressing ourselves primarily to the notion of the nation being the final element toward which everything has been developing and after which nothing further will evolve.

We all are still writing about politics. Only the *polis* has become much more comprehensive and cannot be divorced from the social processes. The political history of the last century could clearly be divided into certain fields—diplomatic history and so on. Some special fields like economic history were outside political history. If we look at much of our talk on social history, we still have the divisions which existed in the previous century in mind. But a certain change in the meaning of the term has taken place, and we should be aware of this and should not confuse the earlier with the more recent meaning.

If I may now come to the issue of new methods and tools which we discussed at the end of yesterday afternoon's session, I would like to mention that I grew up in an atmosphere of organized scholarship and my generation rebelled against this system. We tried to get away from this system of hierarchically organized scholarship. At present we are living in the beginning of industrialized scholarship and the possibility exists that everyone will be happier and eagerly industrializing. It is important therefore to consider how to organize and to think through the relevance of these new methods to the various eras of history. Perhaps Professor Stone will be willing to say something about this problem with respect to graduate programs.

STONE

I was rather bothered that after two days of sitting around this table, Sam Beer could ask in a very skeptical manner: Is there a new history? And I think Stephen Graubard replied to him very cogently. There is not an identity crisis in history; there is not a turning away from politics, but a shift, a reorientation of the nature of the problems. Social history is extending the range of institutions recognized as affecting politics. Institutions like the family, the school or university, the police, prisons, mental hospitals—all are recognized as affecting politics.

The other major emphasis in social history stresses social process: social structure and social mobility, conflicts of status and class lying behind the clash of political parties. It is a digging-in and a broadening of the nature of politics and the analysis of social structure.

In terms of intellectual history, my impression is that there is a search for a hidden meaning behind the data of intellectual history, whether it be the meaning of words which Messrs. Gilbert and Pocock are working on, or the psychological undertones which Mr. Manuel is working on, or a fusion of broad cultural forces which Mr. Schorske has been working on. In intellectual history, there is a new, more sophisticated, and subtle attempt to link intellectual history with a social matrix. And, of course, a whole new type of intellectual history has just suddenly exploded—the history of science.

We have also witnessed the development of, for want of a better word, materialist history. Historical demography has perhaps been the most important and fruitful field of historical adventure during the 1950's and '60's. It has certainly been the most successful in terms of concrete results. As a corollary to this, there has been an increasing stress on the old factor of economic history, but with much more sophisticated statistical techniques and theory.

So these are the three main new things that I see happening: social history, intellectual history, materialist history.

One approach we have not talked about at all is what the English call prosopography, or group biography. An enormous amount of work is being done to construct a group portrait of a set of actors—be they Members of Parliament or intellectuals—on the hypothesis that people's mental attitudes are largely determined by their cultural conditioning.

Many graduate students are working on these group biographies in one form or another. Of course, the main danger is that students will become slaves, because one must have a lot of people working to collect the data for a group biography. They are professional slaves. This is a real problem; one of these days they will rise and murder their masters, although maybe not in our time.

The second thing that has happened in terms of approaches which we have not discussed is local history. In the last twenty-five years there has been a tremendous increase of depth studies of local communities—not pious histories of a local community by some local parson, but studies to illuminate the general problems of the nation-state. This approach has been particularly successful in France and in England. To my knowledge, not a great deal has been done in this area in America or elsewhere.

Thirdly, I would suppose that comparative history has developed in terms of understanding the danger of too close attention to the problem of a single nation-state. It has been recognized that a comparative perspective is necessary to identify what is general in an epoch as against what is particular to a nation-state. It is held that unless you have some comparative perspective, you cannot identify what is English and what is nineteenth century. This comparative approach has been particularly significant in studies of modernization and economic growth.

As regards techniques, obviously the most important one is quantification. It is easier to count in the twentieth century than in the tenth, as Felix Gilbert has repeatedly pointed out. The further back you go the worse the problem becomes. Economic history has, perhaps, been most affected by the quantification methodology, but these techniques have spread across the spectrum of history. I do not think these trends are particularly frightening. It is just as easy to be stupid not counting as it is to be stupid counting. One can be trivial with words as well as with numbers. We are now beginning to get data storage banks and group biographical data.

It seems to me that the application of theory has affected enormously the work of the historian: theories of economic growth, of revolution, of modernization. And I am not so gloomy about prospects for the future as others here are. The problem of the graduate student slave is the one exception, and I think we should give it great thought.

LANDES

On the basis of some of my own work, I must say I am rather suspicious of the notion of group biography. I may be wrong and would just like to know more about what is actually being done.

STONE

Such efforts are spreading across the spectrum of history. The two easiest things to find out about a man are his business interests and whom he married. It is all too easy to assume that his behavior was affected by those two facts and nothing else. Namier himself walks straight into this trap because he determined people's political behavior exclusively on these two criteria. This was all he could find. To find out about people's religious opinions, for example, is extremely difficult.

SCHORSKE

May I speak to this? A German historian, Heinz Gollwitzer, wrote a book, *Die Standesherren,* in which he traced a group of the nobility at the close or in the midst of the Napoleonic era. These were the eighty-three families that were mediatized—that is, they did not get back their princely prerogatives after the war. He studied this class in terms of its social role, political behavior, and so on. They spread throughout the Continent and played a tremendous part in Central European politics, especially in the development of aristocratic-liberal parliamentarism. This book is one of the best cases I know of real payoff from group biography.

BENSON

Certainly in respect to political behavior, the analysis of the social composition of political parties in comparison with the members of the House of Commons and the members of the French Assembly or the members of the American Congress is just beginning to be done. People are finding some interesting similarities, but also some interesting differences. In such work, differences in the outputs of the different political systems surely cannot be ignored; nor can the possibilities of the differences in the system.

HOFFMANN

At least for the French case, it would seem to me that any attempt at deriving anything at all from a study of social origins is futile. One can draw a fascinating comparison between the educational background and social origins of the members of the Assembly and those of the American Congress, but this would not reveal much about the differences between the political systems.

BENSON

That may be so, but if one compares the different outlooks of the two groups, one would want to know the different recruitment patterns of the political elite. This is a significant variable and should tell us a good deal about the behavior of political systems.

KAYSEN

As I understand it, archaeologists used the term "prosopography" to describe the cataloguing of the frequency and location of the appearance of names on inscriptions. Of course, one frequently knew nothing about the significance of the name on the inscription.

I should like to address a comment to Stanley Hoffmann. It seems to me that an orderly description of social data is the background for any conceptual structure.

MANUEL

This sort of study began to be done by Merton around 1938. Merton had a list of seventeenth-century scientists and then made an attempt at a sectarian definition of their religious beliefs that was sometimes a bit mechanical. I want to propose the possibility of taking a small sample of about ten scientists from the same period that Merton was discussing, and doing biographies in depth, including what I would call the psychological aspects of their life histories. I am not sure what you would come out with, but this kind of study would doubtless raise questions about the family constellations and early traumatic and creative experiences of the scientists.

This area is totally open, and perhaps the efforts would produce nothing, but I would be quite willing to try this approach on various groups. I do not know whether the historians of science, as a body, would be amenable to this method at the moment. I doubt it—though some individuals might be, because there has been some criticism of the singling out and overemphasis of the religious factor in Merton's thesis.

Many of the scientists of the second half of the seventeenth century left autobiographical sketches of one sort or another, which allow you to broaden your study. I would conduct such an inquiry in terms of inspection and generalization rather than of rigid quantification, because the universe would be difficult to shape.

TILLY

One of the main things that has happened methodologically since World War II has been the application of the simple notion that important collective effects can appear in the accumulated experience of considerable numbers of people who themselves do not experience those collective effects as events. This is certainly true of the work in historical demography, which in fact is normally almost identical to prosopography. It is also true in the study of social mobility, which is again just another version of collective biography. In fact, many of the large innovations that people seem to have brought in with the social science influence are really different versions of collective biography. These innovations consist of treating the experiences of considerable numbers of relatively ordinary people in a uniform fashion and aggregating these experiences to reveal some collective effect which otherwise would not be apparent.

This methodological innovation has opened up to the historian the ideas of people studying the same phenomena in the contemporary situation and conversely. Stephan Thernstrom actually has important things to say about the contemporary study of social mobility. The Spring 1968 issue of *Dædalus* nicely sums up the sense in which historical demography is forcing the contemporary demographers to rethink their models of demographic transition.

SCHORSKE

I should like to make one short remark. The Gollwitzer book is directed toward answering the question: What influence did this particular group of the nobility have on the transformation of Central European politics and political institutions? Some people in this group stand out rather clearly and have definition; others recede to the level of statistics and are included because they represent certain kinds of interests. The latter support the activity of the individuals who emerge out of the matrix and who transform the group identity into a factor in the political process.

The kinds of collective biography to which Tilly has just referred are at a different end of the spectrum. Nobody is fleshed out in those statistics; indeed, they need not be because the question being answered does not involve that. Thus, it seems to me that merely identifying group biography as a tool should not deceive us as to the degree of the range of abstraction and precision that is possible or desirable within the use of a group as a unit of study.

LEVENSON

A sense of distance toward a possible object of investigation is a *sine qua non* for the historian. Contemporary history is difficult in part because our sense of distance is rather minimal. The importance of strangeness in leading one into the whole historical enterprise should not be overlooked: strangeness not only in the sense of the explorer finding people who look different and have different cuisines, but often in the sense of paradox where similar surface details in certain situations cloak very different things.

When historians categorize themselves as being social, economic, or historical, they are really just saying that they choose to attack the historical iceberg from a particular side. This is a valid start, but the successful conclusion of the journey is to map out the whole iceberg.

In the Chinese situation, one could take the long traditional imperial bureaucratic history, which they date from the third century B.C. with *the* first emperor. He is the famous man who burned the books and buried the scholars. Shih Huang Ti is also connected with building the Great Wall. In Chinese terms, it is an insult to say that somebody burned the books and buried the scholars, and thus the Confucianists gave Shih Huang Ti a bad press. This immediately raises a fascinating conundrum because Shih Huang Ti is the first emperor and one sees him as a monster at the start of Chinese history. He founded the Chinese imperial system which happens to be precisely the system in which the Confucianists become established.

Thus, you have a system that brings prosperity and prominence to people who from the start seem to condemn its founder. That the Ch'in empire leads into the whole series of dynasties, which have their Confucian dynastic histories, gets one into a question with useful and informing ramifications for intellectual, social, economic, and political life in China: the persisting tension between bureaucracy and monarchy. It helps one to order all sorts of facts to recognize this attraction-repulsion, this sense of belonging together and yet in various ways of straining against each other.

This approach immediately involves one in comparative history. One cannot, however, be Procrustean and imagine that some particular model of bureaucratic, monarchical, or aristocratic relations in any particular place yields an absolute model for arranging the facts of China. But comparisons do pose a challenge to interpret these facts in some kind of relation. In writing a work called *Confucian China and Its Modern Fate*, I found myself using Leonard Krieger's book on the German idea of freedom in ways that were, I hope, suggestive, but that by no means offered a blueprint which I could then in mechanical fashion overlay with Chinese names instead of Prussian ones.

Thus, to me, the question of distance, of paradox, is important. I did a book about fifteen years ago that came out of my Ph.D. thesis. This book was given a hard time by the first reviewer who handled it. He may well have been at some point a student of the man I was writing about, a Chinese intellectual and political figure who was born in 1873 and died in 1929. In the course of writing about this man, I found myself speaking about certain inconsistencies in his thought and dwelling, with what seemed to the reviewer to be loving attention, on all the elements that did not seem to hang together. I felt it was not just by chance that I, not being Chinese and being not too much younger but still a generation or two later than this man, should notice these inconsistencies.

The reviewer chose to take this book as the effort of a brash young man to score debating points and to vault over the prostrate form of one of his betters. The reviewer, in effect, interpreted it as being filled with the parochialism of the present and my own culture-bound ideas. He took me to be saying that this man could not cut the mustard even though he was rather an impressive intellectual figure in the Chinese world.

My feeling was exactly the opposite: that one starts with both an empirical and an *a priori* commitment that this was a very intelligent man, one of the most sensitive men of his generation. It is precisely for that reason that one finds it uncannily interesting that his thought often broke along the lines of tension. These things had to be explained not because what he said was ridiculous, but because what he said was reasonable.

There is a distinction between the timelessly cogent irrational and the historically reasonable. Why is it reasonable for someone to think thoughts that seem from outside that historical situation not thoroughly rational? That is the historian's question, one that can help you to write history. It can also make you want to develop new techniques to get the information that will answer it. We could do more thinking about these problems of the historical imagination.

GILBERT

I should like to get on with a discussion of comparative history, but would anyone like to add something to Lawrence Stone's list of the various topics in new history before we do so?

SCHORSKE

I would like to bring up one approach that has not quite got into the discussion: the single biography of a person of no consequence. A recent instance of this approach, Otto Brunner's *Adeliges Landleben* on the life of a small Austrian nobleman, shows how extraordinarily illuminating it can be. Happily this was one of those cases where sources make things possible. Brunner's nobleman was an amateur scientist; he was an upper Austrian Protestant fighting for the Catholic emperor in the Thirty Years War; he had scattered land holdings and was concerned with estate management, the best kind of husbandry, and so on, and all his household records survived; he also wrote bucolic poetry. He was in a special position within the nobility because as a Protestant he had certain ties to Calvinist Hungary that were illuminating with respect to the seemingly monolithic Catholic empire.

Microcosmic social history, as a complement to the macrocosmic technique suggested by prosopography, mobility studies, and the like, may produce some highly enlightening results.

GILBERT

If we can go on now to the topic of comparative history on our agenda, I should like to ask David Landes about his work in economic history.

LANDES

Of course, a field like economic history has certain advantages because the variables you talk about have, more often than not, been quantified—particularly since the development of national income accounting. Thus the economic historian has a fair amount of commensurable information. There are problems, however, in comparing the national incomes of different countries, not only because their systems of accounting are different, but also because the equivalences are deceptive. But even allowing for that, comparative economic history is enormously facilitated by a certain homogeneity of subject matter. Nevertheless, I am not so sure that there is any more harmony among economic historians than among people studying other subjects on a comparative basis. In a sense, with all our advantages, we are not much more scientific than anybody else.

The analysis of the unprecedented growth of the postwar European economy, for example, has been a source of great debate. If you were to put a group of economists and economic historians in a room and ask for opinions, you would probably end up with as many opinions as there were people. The increase in knowledge, in the data available, has not yet enabled us to find that happy sense of conviction and understanding that is presumably our ultimate goal.

But for all the limitations of comparative history, I still believe that it is the one lever we have to pry open the secrets of explanation. Things you have neglected suddenly seem much more important and conversely. It is quite clear, however, that the technique has not yet panned out as one would like. From what I see of grant applications to foundations and reactions thereto, I think there is some disenchantment with comparative history, a feeling that it has become a tired slogan. But my own conviction is that this kind of study is not going to stop because of disenchantment. We sense that it is indispensable; you cannot do certain kinds of history without the context comparative history provides.

I would argue that, without abandoning the national context of much of our research and training in history, students should be required as a matter of course to educate themselves in comparable phenomena in other countries so that they have some sense of proportion; many more courses than at present should be organized on analytical-topical rather than geographical-chronological lines.

HERLIHY

The comment I was going to make about the Middle Ages largely reinforces what David Landes has said. There has been a long tradition of comparative history in regard to feudalism, for example. Feudal institutions immediately invite comparisons between Western Europe and Japan or Russia. In my opinion, these comparisons have not been terribly successful, but they have had a limited amount of success by exposing a kind of specious causality. In other words, if you consider Western feudalism all by itself, you tend to interpret it in terms of certain given factors. If you apply those same given factors to a different society, you frequently see that things you considered to be important were not important. But I do not think that this is an authentic comparative history, because the end product is really a better understanding of the institutions of the particular society you are dealing with.

POCOCK

A piece of work I am doing at present has to do with what they call constitutional thought in the sixteenth and seventeenth centuries. I have imputed to Western European thought at that time a certain set of conceptual vocabularies for dealing with the accounts of particular political events. I set up a model involving notions of custom, fortune, providence, prophecy, and one or two others. I worked this out, with considerable trepidation, for sixteenth-century Florence and seventeenth-century England. What you get is, I think, a study of how a common conceptual pattern can be found by examining various thinkers dealing with different sets of problems in different institutional settings.

ROTHMAN

Perhaps part of the problem with comparative history is that it is often operated on a highly simplistic level—a comparison between England and America, or between England and France. It is not national lines that you are, in fact, trying to isolate out, but critical variables. Critical variables in a comparative sense may well be found within the nation: a Protestant city in France versus a Catholic city in France, and so forth. But if we continue to conceive of comparative history exclusively in national terms, we may lose the methodological innovation.

HOFFMANN

But doesn't this depend entirely on what problem you are dealing with? For certain phenomena the national borders and national institutions make a great deal of sense precisely because, as Mr. Pocock points out, one can observe what happens to a common body of ideas when it gets carried through different institutions over time.

ROTHMAN

Yes, of course. My comment was in the nature of a caution lest such studies revert to a level of national character and show very little, ultimately, in a comparative sense.

KRIEGER

I have been wondering what is new about the recent interest in comparative history. The interest in comparative history goes back, to my knowledge, at least as far as Bodin. It is fairly continuous and becomes prominent in historians like Burckhardt and Weber. Obviously historians are today more careful and self-conscious about the way in which they make these comparisons. But can we say that there is anything new beyond that? In terms of the most successful cases that I know, we might be able to say that comparative history has recently been applied best in terms of variants in a common historical process, rather than in terms of a comparison against the formal backdrop of a common institution like feudalism. Where the comparisons are not in terms of a static model, but in terms of an actual historical process—as in David Landes' essay and R. R. Palmer's book—the results have been most successful.

GILBERT

The new development in comparative history really began with Marc Bloch's famous speech.

KAYSEN

I asked myself when Leonard Krieger was talking: "Since when has Max Weber been classed as a historian?" David Landes told me then that Weber did his thesis on the structure of Sicilian agriculture in Roman times. But from my reading of Weber, I would say that he was not a historian at all. That is his virtue. If you are to do comparative studies in a serious way, you have to break the fundamental disciplinary rule: You have to work in a situation in which you are not in control of the original documents. Weber was not a serious Sanskritist; he was not a serious Judaist; he was not a Sinologist. But he wrote books about the religion of China, the religion of India, Christianity, Judaism, and so forth. He wrote what is still, to my way of thinking, the most interesting single work of economic history that anybody has written and this about material over most of which he was not in control. Weber was a man who believed what he read in secondary sources and thought you could do something with it. Now that is what comparative history is really about.

LANDES

Let me tell a little story in this regard. We had, as every department does at one time or another, an argument about a junior appointment. One of the points that was brought up against this fellow was that he had relied for certain things on secondary sources. The debate went on and on over this point until Ben Schwartz made the following comment: "You know I write these monographs with the hope they will be useful to somebody."

LEVENSON

In talking about comparative history, surely the influence of Marxism cannot be neglected. The problem of periodization has interested historians in many different fields; once one gets into it, one is in the midst of comparative history. It was not by accident that the Chinese intellectual revolution of the twentieth century began with a group who called themselves, collectively, the Chinese Renaissance. And they knew what they were trying to compare themselves to. If one studies them, one can see where the analogy breaks down, which is most revealing. Nevertheless, they looked to what they thought they saw in Renaissance Europe—the rise in vernacular literatures, the importance of science, certain attitudes toward antiquity—and believed that perhaps this was happening in China. They thought that this would have the promise of all the bright optimism that renaissance has sometimes suggested of onward and upward.

SCHORSKE

It seems to me that two different kinds of comparisons have been emerging here—one is examining two separated units under a single head or question and discovering their comparabilities; the other seems to be tracing the nature of the movement of an institution or a set of ideas from one medium to another. The new context refracts the idea or institution and accommodates it to itself. Thus, we have, on the one hand, concept clarification or reciprocal illumination and, on the other, an actual dynamic process in which one becomes aware of how a given sector of social experience or attitude evolves as it moves through different cultural media.

POCOCK

By and large, although people in the West are taught about other peoples, they do not tend to think of themselves and their society in terms of its relations with another society. But the rest of the world does. There is, therefore, a historical consciousness developing that is in terms of the interchange of self-images—the image you have of yourself and your society in relation to the image you have of another society—impinging upon the West. Here one is going beyond comparative history as a respectable intellectual exercise and getting into some kind of existential self-consciousness.

SCHORSKE

Haven't Westerners always done that? It seems to me that if one studies eighteenth-century France, a certain group of people in France were defining their concerns and aims and objectives in terms of an ingestion of the English experience.

POCOCK

But the intensities were very much less then.

GILBERT

I am not quite happy about the rejection of Max Weber as a historian because he used chiefly secondary material. It seems to me that he suggested comparisons of institutions within the various European countries and proceeded in the accepted conceptual framework of historical scholarship. The pertinent question of our discussion is the extent to which our concepts have changed this framework. As Professor Hoffmann said, the entire issue of national character needs much more analysis. There is now increasing interest and concern with local institutions, local history, and this makes comparisons on a different level possible. To what extent do our new concerns and methods lead to a refinement of comparative history within or among various civilizations? I should like us also to address the question of whether our understanding of other civilizations gives us new criteria for comparison and makes comparisons more valuable.

KAYSEN

In both training and practice, historians do appear to give more weight than makes sense to the question of dealing with original sources.

LEVENSON

There is, I think, a bit of caricature and anachronism in saying that historians honor only the drones. A historian certainly must be familiar with the sources of his area of study; but there is also the demand that his work have intellectual cogency. The manipulation of stuff out of archives never gets anyone promoted in a good university these days.

KUHN

Whatever sort of work one is going to do afterwards, at some point in one's training one has to have worked very closely with sources and built from these sources. Otherwise one will not know the nature of the work that goes into putting together the narratives that are later going to be your own sources. I would myself be prepared to accept a good deal of Carl Kaysen's general critique of the profession and still say that it is terribly important that the Ph.D. thesis itself have been done from source material or that there must be a major involvement with work from source materials in the course of training.

KAYSEN

That is a different point. Obviously history is manufactured out of data that are the historian's primary equipment. Nevertheless, my impression is that there is some sense, broadly speaking, in which the distribution of the efforts of young historians is badly skewed with respect to what the task of the discipline ought to be.

GRAUBARD

Carl Kaysen has undeniably hit on something that is true. One thinks of what R. R. Palmer has done with his *Age of the Democratic Revolution*. He simply admitted that for certain materials he was dependent upon others; he did not, for example, read Swedish. This is the kind of thing that a man can do when he is established and when his capacities as a historian are unquestioned. Young historians would not have thought to do it. It is not a question of whether these younger men would have been permitted to do this.

Take another example. We have all read or at least looked at parts of the Barrington Moore book. He is not technically a historian. Everyone who reads it for the section that he knows best finds some inadequacies, but almost everyone would agree that few historians would have attempted that task.

I would say that all of us in this room would have said thirty years ago that certain kinds of contemporary history could not be done. This assertion would have been based upon two presuppositions: One, there was always the assumption that you needed a complete archive, whatever that meant. The complete archive was never available, and the archives needed for diplomatic history were not even open for the modern period. The second assumption would have been that objectivity is attainable for the more remote period, but not for the more modern period. Both of those things are said less frequently today than they were twenty or fifty years ago. In short, I find a significant openness to the doing of contemporary history that I do not think existed when I first started my own studies as a graduate student.

I would like to draw a parallel from what Lawrence Stone said in respect to social history. If I am correct that contemporary history has become more respectable, there is a possibility that many more historians are going to do it. There has been an ecology of historical disciplines, and history has been a profession in which you could expect a certain number of historians to be engaged in ancient studies, others in medieval, modern, and the rest. This was true through the nineteenth century of literature as well; there was no school for studying modern literature until some time in the early-twentieth century.

We are seeing a vast growth of interest in contemporary history, but this does not necessarily mean that other more remote time periods are going to be neglected. We might, however, reflect on the distortion that might arise from the new preferences of students and scholars.

ROTHMAN

I detected at the end of Stephen Graubard's remarks a tone of caution and would like very quickly to relate some experiences over the last two years at Columbia. Entering graduate students insist in droves on doing contemporary history. By this they do not mean history up to 1952, the date chosen by some historians because of the Truman Archives; they talk about contemporary history as post-1950. P.S.-201 in New York and its conflicts are ostensibly material for M.A.'s. We have found ourselves at Columbia involved with trying to give these entering graduate students the intellectual and emotional feeling that history consists in going back behind World War II. This has not been very easy.

HOFFMANN

I do not understand why you would discourage students from doing contemporary history. In one respect in particular it seems to me that the fetishism of documents can be reversed when you come to contemporary political history. If you want the documents, you have to get them while the people who were involved are still alive, because many of them are not going to leave traces. I have been working on the Vichy period in France, and it seems rather clear, not only to me, that if one does not catch a large number of the witnesses while they are here, one is going to miss a great deal. Much of it is not in documents; it is not going to be found in archives when they are opened; much business is conducted by telephone or by private conversation. Later when the archives come out, it will be possible to correct this first reading; one may be able to bring new material. Nevertheless, a great deal will be lost if one does not make a first attempt at dealing with the event while one is still fairly close to it.

STONE

I think this shift to contemporary history is merely part of a trend that covers the whole spectrum of history. For example, two years ago a statistical survey of the history profession was done in the Public Record Office in London. The survey was suppressed by medievalists because they discovered that the further back in time the field got the older the historians were. Almost all the people in their twenties were working on material from 1900 onward.

I do not quite understand why contemporary history is regarded as a thing by itself. It is not a new method; it is not a new problem. The only new thing, apparently, is a type of evidence—oral records. Its reliability, however, I would think to be quite dubious. The Dulles oral history project is a striking example of this.

LANDES

The problem is not that people have poor memories, but that they remember what they want to remember. Memoirs, for example, are looked upon as self-serving documents. I am sure Stanley Hoffmann has had his fill of this historical cliché. He has been talking with people about Vichy, which is still a hot subject, and everyone connected with something so recent and controversial wants to insure for himself a proper place in history. Historians are sometimes naïve, however, in thinking that this self-serving element only affects memoirs. By 1914, for example, the diplomats were creating documents that they could point to later to show that they had, in fact, tried this or done that to keep the peace, so as to put the future adversary in the wrong and set up the kind of account that would look good. Once you introduce that variable, you encounter real difficulty in trying to assess what did in fact happen.

KUHN

I would like to separate the issue of getting documentation while people are still alive from the question of doing the history while people are still alive. There seem to me to be clear and good reasons for trying to get what documentation one can. These reasons are quite independent of the question of whether or not one is going to do the history at this point.

LANDES

You might not be moved to get the documents if you were not doing the history.

KUHN

I agree that there are practical problems of motivation involved, but we ought to manage them to the extent that we can. I speak with some feeling on this issue because I got myself involved about six years ago in directing a project whose object was to gather oral and manuscript records for the history of modern physics and particularly for the history of quantum mechanics. I decided I was also going to try to work my way into doing history of quantum mechanics myself.

From this experience I emerged with a fairly clear sense that one can write the history of quantum mechanics as one would write the history of nineteenth-, eighteenth-, or seventeenth-century physics up to about 1928, but that one is in a very different situation if one tries to deal as a historian with the subsequent line of development. The reason for this is fairly clear. All of the issues that people worked on up to about 1928 are now dead, with one exception. When the problems are still very much alive, the story is not yet done. As you start approaching that range of material, you are in a situation which is significantly different from the situation you were in when you dealt with the earlier period.

It seems to me that at least for a standard sort of history it is terribly important that the historian know both what the situation was like at point A and also what the situation was like at point B when he starts talking about the development from A to B. You are engaged in a different sort of enterprise if you are dealing with a development which has not yet come out the other end.

HOFFMANN

May I just disagree totally with this?

KUHN

I would be disappointed if you did not.

HOFFMANN

It seems to me that when one is dealing with much of intellectual and political history, one never knows where one is. The story is never finished. This is what something like the historiography of the French Revolution reveals. Somebody who writes about the French Revolution has a different vision of it, depending on where he himself stands in time and political opinion. If you are dealing with certain controversial issues, it seems to me to make little difference where you stand because, by the nature of controversy in history, nothing is ever settled. I hate to bring up this notion because it may challenge the myth of objectivity. It helps if the observer writes about something in which he himself has a certain stake and stand. A history written by somebody who has lived in the period he writes about is going to be quite useful in one sense: It will make it possible for him to orient himself along those lines or issues that were important then. The later historian will say from the viewpoint of sixteen or eighteen or nineteen years later that other issues really ought to be examined. But there is no ultimate answer.

KAYSEN

It seems to me that the virtue of contemporary history can be put this way: Here is a social process going on, and because of the nature of our particular society, there is an enormous record. Every radio program, every television program, is taped by law. It is all there and stored. If we try to do contemporary history in a serious way, we are faced with the problem of what is the interesting evidence. What do you throw out? What are the indicators you look at? To an archaeologist no piece of stone is too small to be worth spending a year digging up, because he has so few pieces of stone. If you took that attitude toward the material from the daily press, you would drown in data. The practice of contemporary history, therefore, confronts the historian with a conceptual problem: What evidence does he want?

KUHN

If I could respond to Stanley Hoffmann, but also to Carl Kaysen. I am not saying that historians should not work on the contemporary period. I am, however, opposed to the implication explicit in Lawrence Stone's remark that there is no difference between doing contemporary history and doing history of the past.

SCHORSKE

I would like to strengthen this point—and also argue against Stanley Hoffmann—by suggesting that the B is the first thing a historian picks out for the determination of a historical problem. The determination of the A, the beginning of the analytic procedure in a temporal sense, is most difficult. If the problem is very contemporary, this issue has been conceptually solved before the analysis begins. From this point of view, I find no charms in the notion that everything flows and that all history will go on forever, and we are all in an endless process. We think history backwards, even though we write it forwards in a deceptive way. We think our way back to how a certain process started and then act as if the whole thing went from A to B.

MANUEL

I also want to mention a form of contemporary history: the history of the future. This is growing all over the world and requires a specialized methodology that is worthy of consideration by the historical profession.

KRIEGER

May I just point out some of the connections, as they strike me, between the things we have been talking about for the past couple of days and the subject of contemporary history? In the first place, the use of concepts from social sciences in doing contemporary history obviously no longer involves the problems of anachronism that were raised with respect to applying these methods to the more distant past. Moreover, many of the historical problems in contemporary history can be handled by historians only by using the disciplines that have been devised to handle those problems in contemporary society. There is therefore a necessity, not simply a desirability, to use these methods. Finally, I would like to point out that for contemporary history the social sciences are historical subjects, because the social sciences are an important part of the intellectual life of this century.

LANDES

I just want to put into the record a point about the significance of photography, one new source of information that has not yet transformed history as much as it should. One has only to think of the time historians have spent trying to reconstruct what the mob scenes of the past looked like to realize the importance of our photographic records. This is clearly an area where there ought to be more systematic efforts to collect such documents. Historians have been giving little thought to this.

ROTHMAN

I am glad David Landes raised the point of preservation. Historians have been very slow in this area. I have some friends who have been trying to work on the riots of the 'sixties. For every ten seconds on the television screen, there are at least five hours of shooting, and the five hours are saved as well as what is screened. It is exceptionally difficult to use any of this material—not because the television people are nasty, but because the shelving is wild and completely unorganized. No one has made the effort to organize these critical kinds of data.

SCHORSKE

You ought to raise this issue with the American Historical Association. They could get the National Archives to put the heat on the television industry.

LEVENSON

To return to the question Stanley Hoffmann raised as to why the recent past is not quite so open to the historian's curiosity as something from the more distant past. There is a question of process involved here. There is absolutely no difference between writing about the day Kennedy died and the day Lincoln died.

KUHN

That is clearly wrong.

LEVENSON

I do not think it is so wrong. This question raises the issue of different arcs of time. No historian would want to leave out a history of World War I just because it lasted only four years. People do and should write such histories. Also, I think it is not wrong for people to write a history of parliamentary developments that covers several centuries. The question is somehow different if you are going to go from Runnymede on to the nineteenth century. At any point along those six centuries, you have lots of subjects that are of smaller scope and also get the historians' attention. I want to suggest that the more recent work is not to be barred because it is recent, but that one must consider it in terms of a long arc of time that goes back instead of going forward.

GILBERT

Are there any other remarks on contemporary history?

KUHN

On this question of documentation, Carl Kaysen is quite right that there is a frightfully large amount of documentation now available. The existence of this particular kind of datum should not be allowed to obscure, however, the increasing scarcity of other kinds of documentation. I think now particularly of the manuscripts and working notebooks of scientists from the early modern period. People move around so much more now and have so much less storage space that they often throw away all the papers they have accumulated when they move. In the past one would often find a lifetime's correspondence of a scientist. This is now very much harder to find. The use of the telephone has been particularly significant here.

The Languages of
the Humanistic Studies

This was an exceedingly difficult conference to plan and not an easy one to hold together. Memoranda were prepared for the meeting, and many of them were excellent. Because they were constantly referred to during the conference sessions, and because there was no possibility of reproducing them in this volume, the presentation, in dialogue form, of any significant part of the transcript was virtually excluded. The Editors thought first of simply reproducing individual interventions, not making any particular effort to create continuity. That suggestion led, quite naturally, to the idea that the interventions of only a few members should be reproduced, and that they should be assembled in such a way as to make it possible for the reader to have the total contribution of a single member at once. Such a presentation would obviously destroy the quality of "dialogue," but yield other advantages. We decided to experiment with it.

Participants
Clifford Geertz
Martin Malia
Talcott Parsons
Roger Revelle
Eric Weil

CLIFFORD GEERTZ: On the Languages of Humanistic Studies

This whole discussion of what the humanities are sounds old-fashioned to a social scientist. We stopped discussing where sociology left off and anthropology began and what the social sciences "are" a long time ago. And I would guess that to try to talk about definition independently of what turns on the definition is pointless. Somehow or other, we define things or subjects—humanities, Romanticism, art—in terms of some purpose which we have in mind. And we choose different kinds of definitional strategies and therefore different kinds of definitions depending upon what it is we are up to.

I just wrote down four examples, and there are fifty, I am sure. You could approach a definition of the humanities, for example, with a kind of pure-laboratory-preparation sort of notion of what is criterial of the humanities—uniqueness, subjectivity, and that sort of thing. And this would have certain uses; for example, if your concern is with philosophical issues—that is, if you are trying to investigate what uniqueness means. And it would also have certain uses—I do not say this in any negative way—ideologically as contra-science definition.

Another kind of approach, which has also been suggested and apparently was the one involved in John Higham's article, is a kind of extreme nominalism: that is, the humanities are sort of what people generally say they are. This is useful for academic divisions, for inviting people to conferences, and that is about as far as you need to take it for such purposes.

A third approach that has been suggested is an *ad hoc*, case-by-case inductive approach: You decide what the humanities are by looking at what people who have said they are humanists are doing, and see whether this coheres into any kind of a recognizable intellectual object. This obviously is useful for the self-image of the people working in the field, for defining the problems that have existed in the field, and so on. You can also talk about the interplay among disciplines as defining the humanities—what poetry, art, history, music, and so on have to give to one another. Thus, you can define the humanities in terms of interdisciplinary process; by the way in which they are involved with one another and with the social sciences, in terms of the inputs they contribute to one another and the outputs they yield, in terms of the kinds of relationships to one another they have. This strategy would define the humanities as a kind of unity in terms of the sorts of relationships specific studies have among themselves.

Now I realize I have given only four approaches among many possible ones. But I think to talk about defining the humanities independently of purposes and ends is really not a very good idea.

Someone asked me at the end of this afternoon's meeting how this discussion differed from others that, say, the social scientists have, and I said I was struck, as an outsider, by the lack of any reference, in attempting to define the humanities, to what was going on in the fields of the people represented here. There has been a good deal of discussion of what the humanities are, but there has not been much discussion about what the movements of thought in literary criticism or in history and so on are. And I was thinking that an example of what I mean by the way a social scientist would more likely go at the same thing is the paper by the psychologist, William McGuire, in which he tried to show what was going on in social experimental psychology. I do not entirely agree with his paper, but he says they are moving people out of the laboratory into dealing with the kind of material humanists use. It would seem to me that *maybe* some people in other specialized fields represented here could do some of the same sorts of things to show which way things are in fact moving in the humanities or try to define the "humanities" they are involved in so that we have some notion about what is really at issue.

I think it is quite wrong to say that there is no audience for the humanities, whatever they might turn out to be. It is not true that only humanists read and look at paintings and so on. There is a very large intellectual community in this country. When first-rate works of the humanities are published and written, whether they be Gombrich's *Art and Illusion* or some of Frye's work, I think there is a public response and a larger public than is usual for this kind of work.

The image of our society as a simple machine is a very undifferentiated and, if I may say so, from the sociological point of view a very simple-minded one. As a matter of fact, the humanists do have a function in society which is recognized and, when performed, appreciated. People do respond to you—not quite so much as they respond to the columns in the daily newspaper, but I do not think that is any reason for despair. They probably read you more than they do sociologists. I think at least one major function of the humanists is to lead people into the work, and lots of people want to be led into the work. Wallace Stevens is not that accessible to the average man, even the average educated man, but lots of people read him and would like to get further into him.

One of the reasons why the discussion is centrifugal is because every time we come upon a methodological issue we avoid it like the plague. We raise a fascinating question—that Keats can be treated effectively in synchronic terms, as a presentation of work, while Blake can only be treated in temporal terms. And then we say, this is only an empirical problem. This is exactly where the methodological investigation (and by this I only mean being reflective about what you are doing; I do not mean creating jargon, though we can do that if we wish), exactly where the methodological problem, should begin. What is the difference between Keats and Blake that they should respond to the critical eye so differently? This is the kind of question, it seems to me, we ought to begin with, instead of passing it over and saying it is just an empirical problem, and we all know the difference. *I* do not know the difference, and I would like to think further on it.

If you start thinking of what the problem between Keats and Blake means, you are going to get into methodological problems. And one of these is what "consistency" means. One has to begin to think about how one determines "consistency" in a poet. What does this mean? Otherwise the whole proposition is just a tautology. I do not know anything about Keats and Blake on this level, so I am not trying to make any substantive contribution to this discussion. I am only saying that when we come across something of this sort, this is the place to start a methodological discussion.

A point on language. I think the pebble theory is extraordinarily simple-minded to be held by people who are supposed to be the custodians of language. Take a word which went through all the memoranda, "subjective." Even as an ordinary language term, "subjective" means so many things. Sometimes it means private as against public behavior, distorted as against objective in the sense of neutral balance. Sometimes it means phenomenological inside views as against public, outside, stereotyped views. Sometimes it means emotional as against detached. You use words like "subjective," and you do not unpack the concepts implicit in them; you do not explicate them. Again, this is where methodological issues, in a very commonsensical way—nothing social scientific or jargonish about it at all—ought to begin. Otherwise, you say the arts are subjective and you do not know, and we do not know, what is meant.

One aspect of the concept of "deep structure" is the difference between competence and performance. This is what makes it more than merely logical. The control of the base structures, which are neutral, is considered part of the basic linguistic competence of the human animal; and this is different from speech which is performance. Therefore "deep structure" is not just a logical structure to account for things observed. It is an argument that these competences are actually wired in; they are given. You can produce sentences according to these rules; you can make empirical rules.

CLIFFORD GEERTZ: On the Languages of the Humanistic Studies

Primitive societies are full of critics. It is not written stuff, but the notion that criticism did not exist before literacy is nonsense.

Is there a different way in which things are retained? I suggest that the step diagram of the synchronic present might be drawn as a spiral for the humanities because they proceed by achieving a very partial recovery of the past and then surging forward. That particular surge is exhausted, and there is another recoil to the past. This is a very selective thing. The New Critics rediscover the metaphysical poets and neglect the Victorians, and this sudden burst of new energy then exhausts itself and people rediscover the Victorians. You get this kind of surging back and forth. Using now the psychologists' difference between short- and long-term memory, by definition, the humanities (as Professor Weil suggests) have to have an elephantine memory and must never forget anything; they must keep open the possibility for this recoil to any material which may become relevant. Now obviously this is not possible in a total sense. But there has to be some way, even when the Victorians are not being read and are not relevant to what is really going on—as they were not so much in the high tide of the New Critics, for example—to preserve their accessibility. There must be somebody to keep them alive for the time when this particular surge forward is exhausted, and it is necessary therefore to return to them. So one must make the distinction in the humanities—and in the more humanistic humanities the more so—between short-term memory which is extremely selective, and in which I think in fact it is necessary to forget a great deal in order to make some advances, and long-term memory where it is necessary never to forget anything. You keep the whole enterprise going so that this particular surge will not be the last one.

In the sciences—again, the more scientific sciences, the more so—memory tends to be embodied totally in what happened yesterday. We do not forget Newton, that is quite wrong; Newton is in some way embraced in more recent things because they are built on Newton's suppositions. The reason our students can accumulate it all is that somehow or other it is embodied in short-term memory. You do not need the long-term memory because there is no need, in order to be a beneficiary of Newton's discoveries, to read Newton, in the way in which Mr. Hutchins, at one time, thought we ought to go back and read the basic text. This is not necessary because somehow the latest physics textbook in some sense of the term embodies that advance.

Most people I know who have theories in the sciences (and I would imagine most people in the humanities too) are in favor of them and try to marshall as much evidence as possible to support them. In time the theories are eliminated as people criticize them; but the actual orientation most people have toward their work is to try to support it. Hypotheses in the sciences are always under evidence; they are always under inspection. To look at the way in which people approach their work in either the sciences or the humanities with this sort of negative notion—to say that what they are really trying to do is disrupt the sand castles they are building—strikes me as empirically wrong. They fight for their castles, as we all know, with great vigor.

There is a particularly telling case in Gombrich's treatment of Constable. The usual interpretation of Constable is that he just looked out there and saw the trees and painted them and so on, but Gombrich tries to show that Constable is really better understandable from other painting traditions on which Constable drew than from how the English countryside "looked" in some trans-artistic sense.

Gombrich has trouble at both ends when he makes the distinction between illusionistic and iconic art, because when you are talking about primitive art it is hard to know what the image is being matched to. And it seems to me that this is a good example of how progress develops—you can see here where you might go next. What obviously needs to be expanded now is the concept of matching. The making side, the codes and so on, he has straight. But he has a rather narrow and, I think, culture-bound view of what matching is.

I think that if the concept of matching in art were explicated more, if it were broadened, if it were clarified, if the post-Gombrich sort of tradition were to be turned in the direction of evolving or developing a concept of matching which will preserve Gombrich's tremendous insight into art and this coding business in such a way that we would not be left helpless before non-illusionistic art (either iconic art, on the one hand, or post-impressionist art, on the other), we would see substantial progress. This is the sort of thing where even if we are talking about advance we can see where the problems are left by a thinker of tremendous power. Faced with Jackson Pollack, on the one hand, or primitive art, on the other, something needs to be done to carry this very valuable tradition forward and to make it workable for something rather wider than the particular domain—Western illusionistic art as distinct from Chinese illusionistic art—to which Gombrich has confined it.

In another essay—the essay *Meditations on a Hobby Horse*—Gombrich has a child who, he says, starts with a stick that looks like a horse to the child. It does not look like a horse; but it is a horse to the child, because he can ride it. This is why the Wittgenstein II thing comes to mind. So he rides it, he runs up and down the room, and after a while it is not quite satisfactory to him, so he puts an eye on it, and he puts a tail on it, and so on. So abstraction proceeds in this view from the most generalized representation toward more and more matching as you make the situation; then you match it to the horse. It is not really true that this is a Hegelian view of the interplay of forms at all. It is a theory about the way in which abstraction moves: that is to say, from abstract conceptualizations that are making things—icons—toward realism, toward imagery, toward figuralization, which is the matching side of the question, rather than the other way around. This is quite contra-Aristotelian—they cannot both be right. I am not here to say who is right, but they cannot both be. In the case of the hobbyhorse—which is just a metaphor here obviously—either the child starts trying to make the stick look like a horse and then abstracts it, or he starts with the stick and tries to make it more like a horse. It is logically impossible to go both ways.

As to this humanistic notion of balance in taking care of all the texts, I am not so sure that it really stands up. If it is real oil, then I do not even mind misreading. I do not want to try to defend Leslie Fiedler, but I might defend D. H. Lawrence's readings of American literature which are rather cranky in some ways, but are also extraordinarily perceptive. As distortion plays an effect in art, so I think it plays an effect in criticism, and the notion that only a balanced judgment, only one which will take care of all the text, only one which will meet scientific canons is a valid one is, I think, a somewhat dubious proposition. Some of the best insights we have from "humanists" into the artistic and literary tradition have come from guys who were half-crocked, and this, I think, is something that has to be taken into account. This having to make sure an interpretation accounts for all kinds of things strikes me as eliminating some of the really powerful ideas we have. So I would say that if it is real art, which is a good question, then it is real oil, and it does not matter too much whether some parts of the poem or some other parts of the text will not yield to a particular view. You simply have to try to forget about them.

It is very necessary to point out that we should not assimilate anyone who makes a contrast between deep and surface structure or who is a-historical to structuralism, because the thing that is characteristic of Lévi-Strauss and Chomsky is that they are arguing for deep, pure, formal structures. Others who make deep-surface analyses—Freudians, Jungians, mythological

CLIFFORD GEERTZ: critics like Frye, whom I would not call a structuralist—are looking for content. People like Lévi-Strauss are arguing for pure, syntactic forms, deep structures in the pure sense without interpretive content. Lévi-Strauss is looking for deep binary oppositions, which are supposedly universal, and so on and so on. These people are formalists as well as analysts of the contrast between deep and surface structure. There has been a long tradition, of course, of the a-historical approach: the Jungian tradition, the Freudian tradition. And in anthropology there are a whole lot of people who are looking for what you might call a content substrata, and it is a quite different thing. I do not say they have nothing to do with one another, but I do think that if structuralism in a somewhat narrow sense is to mean anything, it must be this notion of logical transformations between surface equivalents from which you are then allowed to derive neutral base structures. If you do not have this, then structuralist means anybody who is not a sheer empiricist.

I do not agree that Lévi-Strauss does not have an ontology, but that is a factual argument about Lévi-Strauss. But there is no doubt that Chomsky does. If we accept generative grammar, must we accept the return of the nativist? Must we have innate ideas? Do we have to have all this jazz? Having committed ourselves to generative grammar, we are committed then to this kind of rationalism.

Chomsky and the Chomskyites have turned to two things: One is childhood acquisition of language and the other is asking informants whether sentences are well-formed, whether they are grammatical. I. A. Richards began to do this sort of experimental work years ago by asking people about accepted poems: What they thought they meant, whether they "liked" them and why, and so on. Is there anything going on in the humanities these days where people are trying to see how people actually react? I am not talking about sitting here and reasoning how they react to *Hamlet*— but about asking kids how they respond to simpler poems or what people will accept as a poem or a drama. If humanists are going to approach the social sciences and get anything from them, this is one direction in which I would say they ought to move.

Henry Nash Smith said earlier that the humanists are ex-clergy, and he seems to want to restore them to that role. I am unwilling, personally, to accept them in such a role. I do not think they are moralists any more than anybody else. What they can do, as I think Hillis Miller and others are trying to say, is to get us into the documents where these values that you supposedly want lie, and they are hoping to make these documents accessible to us, to get us into them, to find the values that are there. If they get us into *Mrs. Dalloway*, we may find vacuity at the center, but the point is we will find whatever is there. If there is something in Wallace Stevens, if there is something in Milton, whatever it is, their job is not, it seems to me, to be secular priests of our society; it is somehow to get us into contact with the luminosity of the documents, to see the values they contain—and I think there are values, of course—so that we can get out of the poems the values that are in them in this rather incorporeal sense. I was talking to Arnold Stein about Milton and *Othello* before the break, and I learned a lot about what Satan means and about the whole business of the moral career of the self-deceptive man. This is what we want from humanists. We want some way to get at whatever this treasury consists of. I do not think that people who are trying to devise methods to get us into the text are abdicating a moral role. Indeed, I think that is their moral role, because frankly, without offending anybody, I would accept no one here as a moral censor on my life. I might accept Shakespeare, but I am not likely to accept humanists as such.

A "professional" comment about luminosity and some of the problems about simply bracketing so completely the object of study may be useful here. As an anthropologist I am likely to be on the side of cultural conferral of meaning on things. If you look, as I have done, at another society where the luminous objects you are looking at are not luminous for you, you have a lot of digging to do on a number of levels. I have done comparative work on Islam in two countries. Obviously the texts and so on that these people respond to are not luminous to me, so my initial empirical problem is to find out what they respond to and what they see in them. And you cannot do this, if you are outside your own tradition, simply by sticking to the text; it is utterly impossible—take my word for it. If you do not take a historical picture of how they grew up, if you do not have some psychological analysis, if you do not have some notion of their value system, you just will not get anywhere. You will just make it all up out of your own head. You will transfer your own concepts to your informants.

I think the same thing is true, though it is hidden, when you work in your own tradition; and therefore, though I am willing to start off with a notion that there are luminous things—because I know that there are, that there are for me—I am not willing to stop there. I think there is a slight fear, which I do not quite understand, at least on the part of the phenomenologists, that if we begin to subject a privileged document to ordinary explications—if we subject it to history, if we subject it to linguistic analysis, if we subject it to a host of other things—the luminosity will somehow disappear. This shows rather a lack of faith in what is supposed to be the intrinsic quality of the document. I do not think this is true. This is the sort of thing that used to worry Biblical critics a great deal—that as soon as you looked into the Bible, its power would be destroyed. Of course to a certain extent this happened, but to a certain extent it did not. If the object is really luminous, I should not think that knowing something about it beyond its own value will necessarily destroy its hold on you. What does destroy its power and its hold on you, from my point of view, are certain social changes, and this issue is something that we perhaps might talk about.

One comment on privileged documents. Meyer Abrams said he does not mind problems, but he does not like mysteries. In the same way, I do not mind privilege, but I will not stand for undefended privilege. If I disagree with someone over the intrinsic value of a document, I cannot be persuaded to understanding by such comments as: You are wrong, or you are insensitive, or you are an ape looking in a mirror. If somebody is going to say, "This is great," and I think it is awful, I am willing to be convinced. But I need to be persuaded by something besides an entirely intrinsic kind of argument into accepting as valuable something I think does not come to much.

Just a little remark about the theme I used in making my remarks about Lawrence. If you take the view that I have been taking, that the role of the humanist is to bring people into the work, you must realize that it makes the humanist enterprise a very dangerous business. I did not find the fact that your students liked *Mrs. Dalloway* encouraging; indeed, it shows that they have a good sense of nullity, a feeling for the nihilism and the vacuity and despair which are at the center of *Dalloway*. But if you talk about Cèline's writings or *Pisan Cantos* or a lot of other things which are terribly powerful imaginative works, you are not exactly recommending to them the good and moral life. You are putting them in contact with a document of extraordinary power, which is also of extraordinary corruptive power. You are not just playing around with the good little values of the middle class. You are confronting what man has done and expressed about himself both negatively and positively, and you are trusting, I suppose, in some faith that ultimately people who have access to this will come out better than they were before—though I would not like to guarantee it. I do not think you can take the view that the humanities are a kind of substitute religion even at this level, because I think if you are going to teach Cèline, you are going to be in trouble; there will be a lot of people who will believe him. If you allow people to read Cèline and you help make them read him better than they read him before, you are doing dangerous things.

TALCOTT PARSONS: On the Languages of the Humanistic Studies

I should like to make two distinctions that seem to me to be very important. In all the discussion about uniqueness and generality, one should not apply this distinction only to classify concrete phenomena. The tree is unique, but as a member of a species and *qua* member of a species, it is also not unique; it can be classified. I do not think any intellectual discipline can do one *or* the other. It has always got to be doing both in different connections, but in application to the *same* concrete references. I have never read a critic's discussion of a poem that ever really claimed that the poem could not be classified in any way whatever.

The other distinction I want to make is that the treatment of an entity as unique—for example, from the point of view of its development—is not the same as a denial that process is orderly. It is not saying it cannot be conceptualized. Take what Erik Erikson did with Luther. His account of Luther's personality development is not saying that the process cannot be conceptualized; it is not saying that here is a series of events that just happened, without order. He is conceptualizing. It seems to me that to say that if something is unique, it is not orderly, is to make an utterly false dichotomy.

The distinction between the relation of the physical scientist to *his* object and that of the humanist to his is vital indeed. You might say that the humanist's objects—and I think there is good reason for emphasizing documents because documents are not living actors as such—can be abstracted from the nexus of relationships in which the scholar is involved and immersed. Sunsets do not punish observers of them or reward them in the action sense. They may be judged to be beautiful or they may be negatively valued, but what the student does does not provoke a reaction on the part of the sunset. And what the student of a Shakespeare play does may resonate within him: It may activate all kinds of motives and associations, but the play does not do anything to him in an action sense.

The student of a social system, however, is almost by definition part of it, unless he is studying past social systems, which is one of the reasons why history, I think, *is* humanistic; but what the social scientist *does* affects the system. Of course, this happens in many of the physical sciences too. The social scientist provokes reactions in the system which are motivationally needed. A great deal of the methodological worry of social scientists concerns interview problems; what does the way subjects are approached by the interviewer do to the kinds of data that the interviewer gets? It is an interaction relationship, not a one-way relation. Now social systems (at least a large category of them) are, as we sometimes say, institutionalized embodiments of meaning which is of the same basic order as the meaning studied by humanists, the meaning of symbolic systems. This subtle distinction is often very difficult to keep straight. Let me give one example. A set of rules, whether it be at the grand level of the constitution and its interpretations or at the level of the operational rules of some workingman, can be studied in the actual operations of social interaction processes—how they are contingent on the interaction. The most obvious case is that of the law. You can take the books in which the laws are stated and study them as documents. In fact, knowing what the documents say is an indispensable ingredient for understanding the actions in which they operate. Even when the action does not comply with the rules, the rules are there as a critical part of the system you are studying. This is a third relationship that is not comprised either within the natural sciences or humanistic studies.

The very central point about the importance of critical mutual studies of one another's work presupposes certain standards, however vaguely or loosely they are defined, of what is in and out of court, of what is legitimate criticism and what is not. As a first attempt, I thought of two primary components of those standards. One is something that we could very broadly call competence—that is, it is legitimate to expose incompetent work, work that could and should have been done better. There are, in fact, standards of how it ought to be done. And the other is the general area of professional ethics, which may be called integrity. In other words, if a scholar is putting over a hoax on his readers or is playing fast and loose with the evidence, he can be justly criticized. Now probably this pair of standards can be rounded out, and each of them needs much fuller definition for any particular field. I think these are analogous to legal standards that courts will use to judge cases, but in each subfield of the law, they have to be specified; the most general formulations will not do.

I would like to suggest a filling in of the sociological background of what Mr. Geertz said. There has been a tremendous upsurge of the intellectual life, to the great surprise of most of the interpreters of the trend of modern societies, as those interpretations crystallized in the late-nineteenth century. The most massive institutional manifestation of that upsurge is higher education, including research. In the great capitalism-socialism debates, higher education was never mentioned as having any bearing on the future of society—it was the form of organization that was almost exclusively crucial. In the face of this upsurge, I think, it is patently wrong to say that the great humanistic traditions are simply a fifth wheel. The continuity of these developments with the main cultural traditions of society is, of course, rooted in religious history, but also in the Renaissance and all the great cultural movements. It seems to me, the humanistic disciplines are the contemporary bearers of the development of this crucial aspect of these great traditions, and they are integrally central to the society. The sciences are another branch, but they cannot stand alone for a great many reasons, and I include my own set of disciplines in the sciences.

It is quite clear why authenticity of text matters; you cannot raise questions of meaning without relating text and context. If you are going to ask what an author was meaning to do, you have got to ask who the author was, the situation in which he lived and wrote, and so on. To say that you do not have to worry about fragments of writing—how they fit together, where they came from, under what circumstances they were written, and so on—is to limit yourself drastically. In this respect all three of the main categories of disciplines have a factor in common; they are all concerned with the accuracy of description of empirical phenomena and the context within which these occur and never only with the phenomena of maximum salience. This is true of the natural sciences and of the social sciences. History, it seems to me, represents an important case which is genuinely a member of the humanities and the social sciences at the same time.

We have seen in the last generation or so a very important movement of the concern of historians with what might be called the relevance of social science to their work. Of course, not only social science but geography and climatology and heaven knows how many other things are concerned with this. The social sciences are more crucial for the historian.

The facts of the social sciences are always changing. But this process of change is, in part, the product of interaction between the work of men of thought and men of action. There is a parallel here in the creative artist's dependence on the critic and vice versa. The distinction between events and textual objects is a relative one. Events of human history and social phenomena generally are considered by the social sciences to be the outcome of a continuing subtle interplay of the ideas and plans of individuals and groups and the situational and other factors of the processes to which their actions are subjected. Therefore, in some sense, culturally, idea-wise, and so on, historical events are formed. They are not formed in quite the same sense or on quite the same level as works of art, but to say that they are exactly like geological events, independent of human intervention by mining or something like that, is not accurate. In order to understand the event, we have to know what people in the historical background of the events thought, what they were trying to do, and so on. The theoretical social sciences *vis à vis* history are attempts to systematize our more abstract knowledge of the factors that enter into the historical process.

What different kinds of standards of advance are relevant in different connections in different disciplines? A goal stands at one end of the continuum, as if a committee of the American Council of Learned Societies formulated a goal for the humanities and set up a five-year plan by which everyone would be able to calculate how we were doing in reference to that goal. Quite clearly, the main process of advance is not of this character. Ours is the type of society where the main trend of development—looking at it as a social scientist —is evolutionary. Certainly there is change, not random but directional change. Although there is a lot of planned activity at the sublevels, the main pattern of growth and development is not planned. I think the Soviet five-year-plan image is uncongenial in America and should be, though we are quite ambivalent about it and feel rather guilty that we do not have clearly defined national goals as a total society. I think we ought not to. But at any rate, my own view would be that from the point of view of conceptual models the advance of the humanities is much more in the evolutionary direction than it is in the planned transformation direction.

I should like to ask whether deep structure is anything more than the logical fundamentals of any theoretical scheme which has to have more or less primitive terms that are not empirically demonstrated. In other words, the structure of logical systems always has certain characteristics. If you attempt to approach any body of phenomena as objects for theoretical analysis, you always have to get into something of that sort.

I noticed in the memoranda a good deal of concern over whether to emphasize subjectivity or objects. This has become for general theoretical purposes a false dilemma, as developments in social psychology make clear. One of its main trends goes back to James for its inception, but especially to George Herbert Mead who developed it very much further. But this concern also converges with some of the trends in Freud's thought and in Durkheim's, coming from different intellectual roots. The basic outcome is the conception that subject-to-object interaction is from the mental to both the constitution of social systems and the human personality and that basic structural constituents of every personality are, as Freud put it, the precipitates of previous object relationships. It seems to me that this is part of a set of fundamental principles or laws that are basic to human behavior, including thought in the category of behavior, at the level where symbolic processes and symbolic code systems, of which language is a prototype, are operating. There are many antecedents of this, but the human level of linguistically expressed and guided behavior is new in the evolutionary process, and there is a basic continuity, the broad structure of which is coming to be understood much better than before, between the phenomena of linguistic learning and use and so on and the structure of both personality and society. This is a central general background for the problems in the humanities as well as for those in my discipline.

I hope Roger Revelle did not essentially mean that the humanistic scholar any more than any other scholar should *ipso facto* be a man of action, a politician, and so on. This is what some of our student activists want us to be. This would be the old myth that the philosopher should be a king, and this day has passed, if it ever existed. There has been a fundamental differentiation of roles. The scholar is also a man and a citizen, and he has his responsibilities in that capacity, but they are not the direct outcome of his professional work. The world of the intellectual disciplines is one in which, insofar as we specialize in them, different values are emphasized. We each have our own values; we are not independent of values, but we give primacy to certain values. I think the humanists have more of a relation to values and the social scientists somewhat less, but they share this relation because in studying human documents and human behavior they are studying people's values and how they relate their lives to those values. They cannot avoid their high responsibility to understand as well as they can. Their understanding and what they guide students and other publics to may be subversive as well as edifying. This is not, I think, the primary scholarly role. At least as it seems to be important to deal with values for certain social science purposes, these values should be distinguished from both norms and policies. They are at a rather high level of generality; in other words, how you terminate the war in Vietnam is not a value problem—it is a policy problem. Values lie back of such policies and are implicated in them, but the specialist in determining what are the values need not be an expert in policy. The roles should be clearly distinguished. Norms, I think, lie in between, and the prototypical specialist of norms is the legal scholar. There are other norms than legal, but we do not ordinarily think of the lawyer as a moral prophet. The society we are living in has a serious problem of knowing to whom or what groups of people it should look for moral leadership. The situation is, in part, a consequence of Protestantism and has a certain patterned congruence with it—we do not recognize any particular class as uniquely qualified to tell us what are good judgments of value. But I feel very profoundly that the role of moral leader as such is not the same as the role of any company of scholars in the humanities or any other area.

DÆDALUS DIALOGUES

The distinction between artists or poets and humanists can be drawn quite easily. Humanists see the possibility, often the necessity, of refuting one another. But how can one poet refute another? The most important distinction to be made between the humanities and the sciences is perhaps that the objects of humanistic study act upon the man who is interested in these objects in a way that facts in physics never act upon the physicist. A work of art acts upon the viewer, and objectivity is impossible in the sense this word has in the sciences, where it means experimental or observational verification of hypothetical-deductive systems. I do not see how hypothetical-deductive systems can even be produced in the humanities. That does not mean that there would not be criteria of objectivity in this field; but they would be of a specifically different nature.

Much of the discussion here has made me uneasy. We have been talking as if we were outside the field and looking at it from a high vantage point. Too often, I am afraid, we have been discussing history as though we did not belong to history.

What in the last resort is significant? That question can, I think, be answered quite simply. Significant things are those that affect and shape us. I am afraid of our tendency to speak as though we were always choosing arbitrarily and by accident. We make our choices according to the tradition in which we have been brought up. Of course that does not mean that we must, according to Plato's ideal, always do the same things. But when we say we invent a law, we never do invent; we modify a pre-existing law. Similarly we do not choose cultural values; we are not naked souls dropping down from heaven who go to Macy's to buy decent outfits according to our liking. It is difficult to understand how we could have any preferences at all if that were our situation. In the end, we discover what we are by discussing things whose value seems to us self-evident and luminous. These things quite legitimately appear luminous and self-evident to us *because*

ERIC WEIL: On the Languages of Humanistic Studies

they have shaped us. I have great admiration for George Poulet's work, but the astonishing thing is that Poulet treats in a new way the authors you find in any handbook of literary history. This comment is not at all meant as a criticism; it simply is an illustration of my point.

On the value of criticism I can be brief, I think, by quoting an old, unfortunately deceased, friend of mine who lived in the eighteenth century. Lichtenberg said that certain books are like mirrors: When an ape looks into them, he does not discover a saint. I think this observation is a perfect criticism of literary criticism.

Deep structure has always been a subject for people who did any kind of scientific work. One of the key terms in Mannheim's book on ideology, for example, is "structure." This does not take the sting out of modern structuralism, however, because the old structuralism, which in fact goes back at least to Plato and Aristotle, had no rules for transformations. It was a kind of static structure. Although that statement would immediately have to be qualified, it is also true of literary criticism and even of logic. It is not quite so true of history, where precise rules of transformations and generation have always obtained. We are just beginning to shift over from aesthetics to dynamics in the humanities—a change that represents, in my view, great progress.

Our unfortunate position as humanists is that we are interested exclusively in surface phenomena. We are not primarily interested in deep structure and transformational rules when we speak of tragedy; we are interested first of all in *Hamlet* and Oedipus. This is not to say that structuralism cannot be extremely helpful, but structuralism does seem to lead to a science of history, in the ordinary meaning of the word. From it formal hypothetical-deductory systems may be obtained; but then we are back to what the old historians called "auxiliary sciences." They are a big help, but they cannot do the job. Thus I am slightly apprehensive when terms like "explaining" or "generating" are bandied about. We do not really explain a work of art or a historical event by means of the auxiliary sciences. We may indeed show that there are necessary conditions without which an event could not have happened or a work of art would not have been possible. But that is not explaining the work of art or the historical event: We cannot "generate" a work of art or a historical event; we start with it and then try to put it into a frame that permits us to speak about it. My great fear —or, if I may be quite cynical, my great hope—is that we will not be able to digest a historical event or a work of art just that way. Our language, I think, will never be completely scientific, and common usage in our fields seems to show it, when we say, as we frequently do: "Well, old man, what you are contending is absolutely convincing, but I am sure nobody can prove it." This aspect of *convincing* is characteristic of our fields. The old philosophers aimed to lead normal people toward philosophy. But that end could not be achieved by starting with philosophy; you had to entice people to philosophy with unphilosophical means. I am afraid we have to bring people to scientific treatments of unscientific objects with the help of language that will never become absolutely scientific. The procedures of the auxiliary sciences have, thus, very great value, but only in their several limited fields. It is owing to them that in these fields "scientific" discussion becomes a possibility, even a reality, because they make possible observational and in certain cases (for example, in psychoanalysis or political economy) experimental confirmation of hypothetical-deductive systems. But they have to pay a price which seems to be very high if the humanist has to pay it: They necessarily envisage only one aspect of a reality the totality of which constitutes the humanist's object and task. So we shall always have to go back to the task of convincing; otherwise we just have a science of, say, poetic forms and lose sight of the poet and the poem.

Are we not asking: What is the true view of a tree? But there is no true view of a tree; the tree is the totality of all possible perspectives.

ERIC WEIL:
the Languages
he Humanistic
Studies

The humanities are not simply good for understanding other people; they are good, above all, for understanding ourselves because that is possible only in contact with people who are different from us. It might well be that we in our civilization are so much worried about understanding because our civilization has from its very beginnings acknowledged that people with very different ways of life are human beings just as we ourselves are, which elsewhere was not an admissible fact at all, and that for that reason we ought to understand them. People who do not go out of their own setting consider everything they do as self-evident and natural. Contact with other meanings and other ways of life is about the most important thing for an individual—in the context of our civilization.

As a rule, in the humanities we cannot predict the way we predict in the natural sciences. Although we can of course be wrong, we can predict that oil will be found on a certain spot. If by an unbelievable miracle, quicksilver came up instead of oil, we would be most pleasantly surprised, but the scientist would be disappointed because he was looking for oil and oil only. Now, in the humanities, the most important things are not discoveries, but new questions. We cannot predict what the next question will be. Once a question has been rendered "scientific," predictability, or at least verifiability, sets in and boredom follows. Everybody here could write a sociological-psychoanalytic essay on any author whatsoever. But a new question brings the whole thing alive and realizes one of the infinity of possible aspects and perspectives.

The evolutionary model, so far as I can see, means just enrichment, if I may use everyday language. To go back to an old and nearly forgotten critical argument of the medieval schools, we are using in cases like this the *argumentum insufficientiæ*. We are pointing out, in fact, that what the old man said is quite right, but there is more to it. In the sciences, things do get digested;

they are present, but not in their original form. In the humanities, the things themselves survive without undergoing this process of "digestion," which means that in the fields of the sciences the old becomes, if it survives, a special case in the framework of the new and more general theory. On the contrary, the historian and the humanist in general try to re-discover the phenomenon such as it was, even if that endeavor cannot succeed completely because tradition, this collective memory which furnishes us with our objects, has also transformed what it keeps in its elephantine way.

When we discuss literary criticism, we are talking about language concerning language, which is a third-level thing, and I tend to avoid and be wary of third-level things. If we were to talk about events, at least we would get only to a second level; we would be talking about the historian's language.

I should like to take up a question that is related not only to history, but to the whole subject of our meeting. In what sense can we say that we are doing scientific work? What is the minimum requirement for scientific work? Professor Parsons said yesterday that there is at least the necessity of dialogue or discussion; otherwise we get a series of monologues. Such monologues have appeared rather often in literary history. In general history, scientific work in this sense is less rare. But an important distinction remains between history and science in the narrow sense. In physics there is progress; the things that went before can be forgotten. In history, nothing ever gets forgotten, or at least it ought not to be. What Voltaire said concerning Shakespeare was very wrong, but that he said it is extremely interesting and important in order to understand not Shakespeare, but Voltaire, Voltaire's time, and French history. In that sense, things are never forgotten. When we get enough criticism, we get the "history of criticism"; as soon as there are enough "histories of criticism," we get "histories of the histories of criticism."

In such a situation, the only thing we can do is, to quote Pierre Bayle, to throw out errors. But in order to show that there has been an error, we are obliged to develop certain auxiliary sciences. There is the question whether those auxiliary sciences are taking over and history as such is becoming a cesspool into which is flowing everything that cannot be handled "scientifically."

We can determine whether or not a document is authentic by absolutely scientific means. But there always remains the decisive question of finding out what the document means. There are no objective criteria for distinguishing important facts from unimportant ones. When a historian says he is looking for facts, he is looking only for important facts. These facts are important because he goes to his material with a determined question—a question of his time and very probably, if he is a good and great historian, one that has personal significance. Therefore, the kind of objectivity we find in science is missing. On the other hand, we find out quite easily when the historian's inquiry is *not* objective. When a man suppresses or invents facts, when he contradicts his own principles (for instance, when a materialist historian says events happened because people thought of and were obsessed with the idea that . . .), then one can show that he is mistaken and is using a bad method. I can ask for consistency; that he take into account all the evidence; and that he look for new evidence. These are criteria for a discussion that would be scientific in the wider meaning of the word, but not at all in the narrow sense. Auxiliary sciences are, at least in principle, sciences, but they remain auxiliary. They labor for nothing if they do not work for somebody who knows what to look for.

There remains the problem of the unity of history. Let me take as an instance the career of Napoleon. There have been rather interesting books on the social and economic situation in France and Europe during Napoleon's time as well as on the psychological attitudes of Napoleon himself. Many interpretations have been proposed, but where does the inquiry stop? How do we connect these different interpretations, questions, and interests? That leads us back to ourselves, and we are obliged to ask what we are really looking for—a task at once disquieting and satisfying.

I had the chance of meeting many of the great physicists of the older generation, and the remarkable thing about all of them was their interest in problems they called moral rather than in the problems of physics primarily. They found out that you can do physics, but that you cannot be married as a physicist. As a result, there has been a stream of publications, very often of dubious merit, from scientists on politics, morals, religion, and the world situation. We humanists have a guilty conscience when we look at these outspoken scientists, and the scientific people feel even worse when they look at us. I am illustrating facts that everybody knows, but sometimes well-known facts deserve to be recalled.

Our whole discussion here has become skewed to a certain degree because the paradigm we have all been taking is literature. But literature is a particular case because it is language treating language. If we had taken history or philosophy, the problems would perhaps appear different since we would not then be analyzing a given language with the help of language, as in literature. I am not sure that even the ideal of definitive interpretations of works of art can be maintained. It seems to me a very defensible thesis to say that the number of right interpretations of any work of art cannot be limited *a priori;* it could be precisely that which defines the work of art.

**ER REVELLE:
On the
anguages of
Humanistic
Studies**

I might say a word about the method of science that has been contrasted with the possible methods in the humanities, and particularly about the concept of uniqueness which has been referred to again and again. The fact of the case is that all real things are unique. If you looked at a tree and tried to describe the tree in all its complexity, the number of books you would have to write would be bigger than the tree, and they would therefore be useless for an understanding of what you may want to know about the tree. The key method of science in the usual usage of the word "science" is the method of abstraction. This was first pointed out by Galileo; he said "I abstract." Galileo probably never dropped the two balls from the tower, nor did he roll his ball down the inclined plane and measure how long it took to go from point to point. These were both thought experiments—abstractions of reality. He said *if* you roll a ball down an inclined plane, everybody knows that the farther it goes, the faster it rolls; therefore force determines not velocity, but acceleration. In actual fact, if he had tried to carry out his experiment, the results would have been quite variable because of the crudity of his methods.

I was at another conference yesterday where a quite misguided young man tried to define the boundaries of the continents. He was looking for a scientific basis for a practical objective; namely, what part of the ocean should belong to the United States. And he said this could be decided by using scientific evidence to determine the boundary between continental rocks and oceanic rocks. But he had been grossly misled by the statements of geologists and oceanographers about continental rocks around the edges of the oceans and oceanic rocks in the deep ocean basins. Oceanographers are fond of drawing simplified abstract diagrams that show the continents as big rafts floating in a heavier stuff they call oceanic rock, and this is a very useful abstraction, a way of understanding the relationships among phenomena; but a sharp discontinuity does not actually exist in any particular place. You cannot draw a line on the map and say that here is where the continental platform ends and the true ocean floor begins.

If the key thing about science is the abstraction of simple and repeatable imaginary phenomena from complex and variable realities, when you talk about uniqueness, you are saying, in effect, that you reject this method of science, this method of abstraction. You insist on describing the entire tree in all its complexity; otherwise, you say, you do not have a meaningful statement. Well, in what sense do you not have a meaningful statement? The question that might be of interest here is whether the humanities deal not with uniqueness but with concreteness, with the concrete versus the abstract.

When I asked whether an artist or a painter was a humanist, I was not looking for definitions, but rather for objectives or purposes. What Professor Weil has said and what many others have said has greatly clarified the issue. But we have not yet asked what is the objective of the humanities, and this is in some sense the underlying question. I have an obsession, or at least a strongly held point of view, that there is little difference between the natural scientist and the humanist in their purposes or in their objectives, because in both cases their purpose is understanding. I do not believe that, as Mr. Malia has said, the scientist wants to use the tree. The scientist *qua* scientist is concerned with understanding. The humanist also has the objective of understanding—understanding man. The artist and the poet may likewise have this purpose—to understand themselves and to understand other men; but their methods, I think, are quite different from those of the humanist, the person concerned with humanistic studies. The characteristic thing about the poet, or the painter, is that he arouses instant recognition. What he says cannot be proven; it does not have to be proven. What you say is, "I wish I had said that." Whereas the humanist, like the scientist, must demonstrate the understanding he has arrived at. The difficulty is the one Professor Weil stated: The humanist cannot, being a man, withdraw himself from his study because his purpose is to understand man, and man, of course, includes himself. We talked about the sunset, for example.

ROGER REVELLE: You are not really concerned with the sunset as an object; your concern is with how you react to the sunset. This is part of the job of understanding man.

You said that you have the problem of authenticity, and what the scientist would call this is the problem of importance. Understanding the methodology of science *is* a useful thing to do; there is all the difference in the world between the productive scientist and the non-productive scientist. The basic difference is that the one understands the methodology of science and the other does not. It is something that can be taught. The poor scientist asks a question that cannot be answered, while the good scientist formulates a problem in such a way that the answer resides in the question. Let me give you an example. We had a young graduate student here at Scripps a few years ago who said: "I want to study the biology of kelp" (a plant that grows off this coast). That is not an answerable question. You cannot study the biology of kelp; what you have to ask is a specific question. For example: how much organic matter does a leaf of kelp produce in a day? In the very act of asking that question, you have stated how you are going to answer it, because there are quite clear lines of attack. Moreover, the answer to the question leads to other problems, and it is therefore an important question. If, after you have arrived at the answer to a scientific question, it is a *cul de sac*; if it does not, in fact, open up half a dozen other questions—it is essentially trivial. On the other hand, the scientific problem whose answer opens up a whole new set of problems is the important or authentic problem. Is not this in some sense true in the humanities too? Once you have understood what Camus was trying to say about people, will your understanding raise a lot of other questions or will it be simply a *cul de sac*?

Incidentally, I think I understand in human terms the statement that the difference between the scientist and the object of study is the difference between the humanist and the poet. Graduate students, for example, are among the most studied of all scientific objects, but in their capacity of being objects of study, they are not at the same time being scientists.

Does methodology in the humanistic studies have the characteristic that it evolves, that it changes with time, that it grows or builds on the past? People have been criticizing Shakespeare ever since he walked the stage of the Globe Theatre. Eighteenth-century discussions of Shakespeare are clearly different from those of the twentieth century. But do the twentieth-century people build on what the eighteenth-century people said? I do not know whether they do or not. Another way to ask this question is to draw the analogy to science again. One characteristic thing about science is that there is a continually building house or mansion of science; like our Father's house, it has many rooms, but nevertheless it is a never completed structure. Like a cathedral, it is continually being built on the base of what went on before. I do not know if this is true of humanistic studies. My impression, however, is that if there is an ever-rising cathedral of the humanities, it is really a reflection of other people's cathedrals. In other words, what is written does change with time, but it changes because of discoveries outside humanistic studies.

Modern literary criticism used to be phrased, not many years ago, largely in terms invented by Marx, Darwin, and Freud. They were not humanists in the sense that we are using the term, though perhaps they were in some other sense. They clearly contributed to and built on a previous structure of knowledge, and humanistic scholars have tended to draw from their insights and then revise the ways of looking at literature.

Perhaps one source of the insecurity in humanistic studies is that they are not part of a structure, they are not building on the past, and humanists are not sure that anybody in the future will be able to use what they have written and go further.

There is no question about the constant interplay between science and technology. Science determines technology; technology determines science; and both contribute bricks to the structure. The accelerator, for example, would not have existed without many previous scientific discoveries—for example, Maxwell's wave equations. In some fields, such as mathematics, technology is neglected. Modern mathematicians do not think they are being useful; they are proud of the fact they are not useful. They say, in effect, we are playing games, nothing we now do is useful or is ever going to be useful, so far as we can make out.

The drift of the continents is still an important phenomenon. It is not that people have lost interest in it and taken up an interest in something else; rather they have subsumed this phenomenon in a much larger one—namely, the slow churning motion of the entire mantle of the earth, which gives a mechanism by which continental drift could have taken place and a great deal more understanding about the earth as a living thing.

The hypothesis of continental drift was first stated some fifty years ago by Alfred Wegener. He looked at a map of the South Atlantic and noticed that Africa and South America seemed to fit together. His hypothesis that they had drifted apart received support from some biologists, but was vigorously contested by another group of scientists. After a while, it was pretty much rejected on two grounds: one being its physical impossibility—there were no known forces big enough to move the continents around—and the other being different explanations based on the same biological evidence. Now, the hypothesis of drift is widely accepted again, because it has been swallowed up in another hypothesis—namely, that the mantle of the earth is slowly turning over in great convection cells and the continents are being dragged along by this process. So we have had a thesis and an antithesis and then a synthesis of explanations of a set of phenomena. Is this what also happens in the humanities? Do we have a thesis which turns out to be vigorously opposed by somebody else, with both being swallowed up in the long run in a larger hypothesis?

The importance of a prediction is that you make it before you know about the predicted event, and therefore it has a certain charisma. It is awfully easy to explain any phenomenon after you know about it. Everything you find should be, if your hypothesis is correct, explainable in terms of that hypothesis. But the value of prediction is that you are saying you *will* find a particular set of phenomena that confirm your hypothesis.

Let me give you an example from my own experience. In 1952 we made a series of measurements of the heat flowing through the floor of the Pacific Ocean from the interior of the earth, and it was impossible, we thought, to explain these observations without invoking the idea that the rocks in the mantle are turning over in giant convection cells. If the rocks are convecting, then one should find a pattern to the heat flow—high heat flow from something called the mid-ocean ridge, and low heat flow from the trenches which border the ocean. This was not a future event; the future event was our making the measurements. We went out and made the measurements and indeed found that the heat flow is high under the ridge and low under the trenches. The prediction came true and thereby confirmed the hypothesis, even though the predicted event was something that had been going on all the time.

Just because the event has taken place does not mean that you cannot predict—although you do not predict in quite the same way as you do when you talk about events that have not yet occurred. You predict that you will find particular evidence in the rocks, in the evolution of organisms, or in other characteristics of the earth that will be in accordance with your theoretical formulation. There is an analogy here in looking for oil. You do not predict that oil is going to be there in the sense that the oil is *going* to be there—you predict that the oil *is* there. It either is there or it is not there.

I want to say something a bit different, although I think it is relevant to what we are talking about. Let me preface it by pointing out that the next twenty-five measurements of the heat flow, which we thought confirmed our hypothesis, showed a complicated pattern. If we had made those first, we would have abandoned the hypothesis. That is a marvellous example of what happens when you have too much data—you often fall flat on your face. It took about two hundred more measurements before we recovered the pattern we had obtained with the first dozen.

Coming back to our morning's discussion about history and its predictive power, history does not really predict events. It may predict human behavior in a variety of environmental situations. An event consists of the interaction between the environmental situation and human behavior. Let me give a case in point from the issue of *Dædalus* on historical population studies. There are about half a dozen articles in that issue that deal with seventeenth- and eighteenth-century demographic behavior. It is clear from these studies, which are rather varied in their approach, that during the seventeenth and eighteenth centuries many people in Europe deliberately controlled their fertility. This is contrary to the Malthusian doctrine that the population always increases up to the maximum set by misery and vice. People did not have all the children they were capable of having. They not only married quite late, which had the effect of reducing fertility, but within marriage they had fewer children than they could have had. They did this by means that are not at all clear—probably *coitus interruptus,* perhaps abortion, perhaps infanticide, perhaps various folk contraceptive methods. Nevertheless, in the rural societies of seventeenth- and eighteenth-century England, France, and Spain, there was deliberate fertility control. This observation about human behavior is tremendously important and gives us some ability to predict behavior in rural societies at the present time.

Another observation, which is also important from the standpoint of prediction, is that people in the Netherlands and Japan, and presumably in other places, did not limit their fertility until they were placed under considerable economic stress. In the Netherlands, for example, the population went up quite rapidly until about the middle of the seventeenth century; then it leveled off—and in some provinces declined—after the proportion of people who were so poor that they could not pay the poll tax reached about 35 per cent. In the case of Japan, the population in one region which has been carefully studied went up rapidly in the early part of the Tokugawa period until, in some districts, there were as many as six people per cultivated acre, which meant that they were hard-pressed to provide enough food. Then the population leveled off and in fact declined somewhat. This insight into human fertility behavior may help us understand how the Indians and the Indonesians, the Pakistanis and the Egyptians, and the people of other so-called underdeveloped countries behave at the present time. It does not say that they will start to control their fertility to a greater extent than they are now controlling it after their per-capita income declines to a certain point. It says that, in general, people tend to balance off the value of having children against economic advantages. They like to have children; they tend to have children even if it is against their economic interest.

This raises fundamental questions. What are the uses of the humanities? What role do the humanities play for the non-humanist? It was said this morning that it is awfully hard to get into other people's minds if you do not share their traditions and beliefs. That is why I asked Clifford Geertz about the anthropologist's ability to get into other people's minds. Mr. Malia suggested a partial remedy, however, when he said that historical interpretation progresses by increasing the richness and the diversity of our historical evidence. This is, to me at least, a way of saying that in the absence of sharing the beliefs or being in the tradition oneself, one compensates by building up as complete a picture as possible—a process which helps one get into the other man's mind, his spirit, and his heart.

This is really a humanistic study that depends on his literature, his tradition, his religion, his art—everything about him that can give an insight into what goes on inside his head. In this way, the humanities

ROGER REVELLE:
the Languages
the Humanistic
Studies

have a great potential usefulness. They are an alternative means of finding out what people are like, which is at least a supplement and may be a powerful alternative to the tools of the sociologist and the economist. And so we come back to our question of prediction. Would a good humanistic hypothesis be one that would enable you to predict how people will behave, subsuming among their behavioral patterns how they will read a poem, how they will respond to a piece of literature, what their hierarchy of values is, what they think is important, what they think is most important?

Mr. Malia may be saying that the humanities asks questions that cannot be answered.

It took four hundred years of a lot of philosophical thinking, beginning with Thomas Aquinas, before Galileo could think through the problem of force versus velocity and acceleration. Aristotle said that force determines velocity. Thomas Aquinas said if this is so, then what happens if you have no countervailing force —as, for example, in a vacuum? Something in a vacuum would have to fall infinitely fast, because there would be only one force acting upon it. We know that things do not fall infinitely fast—therefore, "Nature abhors a vacuum," even though common sense tells us otherwise. The Schoolmen wrestled with this major problem for four hundred years before it was solved by Galileo. Galileo did not pick out something he could answer from a set of unanswerable questions. The problem had been occupying people since the time of Aristotle.

Perhaps there are two kinds of things involved in this discussion of structuralism. One of them is whether, in fact, the mind follows laws; whether the mind is wired in a certain way. You can ask the question in metaphorical terms: How does the apple know it is about to fall? There may be laws of the way the mind works, just like the law of gravity, that determine the structure of sentences and the groups of words that *are* sentences and those that are not. But another aspect that is possibly not deep structure, but what one might call logic is well illustrated by the book a man called Boole wrote some 115 years ago, *The Laws of Thought*. I believe Bertrand Russell pointed out that it had nothing to do with how people actually think. If it had really dealt with the laws of thought, it was curious that no one had ever thought in such a way before. It was actually a theory of logic by which one could analyze phenomena. It provided the basis of what is now called set theory, one of the great steps forward in logical analysis. By using set theory, one gets away from Euclidian axioms and many other kinds of logical problems that existed prior to the modern development of Boole's ideas.

So it seems to me that one kind of question to ask about a literary work, for example, is whether the writer must follow certain rules because of the way the mind works—just as the apple must always fall. Must the reader appreciate or respond to a work of art in a certain way because of the deep structure—in other words, the laws—of the mind? Similarly, must history follow certain rules because of the laws of human behavior? Now this does not mean that history will always produce the same events; the actual events will be different in each case because each historical situation is unique. If the laws exist, the historical event is an interaction between the laws of human behavior and the particular environmental circumstances. But one could, however, try to find in history a common thread of human behavior—for example, as Mr. Malia suggested yesterday, that liberty going to anarchy produces tyranny. This is in some way a vague and empirical statement of what may be, in fact, a law of human behavior, of the way the mind works. On the other hand, one might analyze history according to a logic that has nothing to do with human behavior, like Boole's set theory, which is a tool of logical analysis and quite unrelated to the deterministic part of what happened, what actually happened and why it had to happen.

I have been listening to the conversation this after-

ROGER REVELLE: the Languages the Humanistic Studies

noon in a more and more disturbed frame of mind. I think you people are in some sense committing intellectual suicide in terms of what your real role is. The average person could not care less about what is a significant document or what is an elite document. The question of whether this is a good poem or not is an interesting subject, but it is an internal subject. The great problems are what is right and what is wrong, how do people act, and can the humanist tell us about how people should act. It is often claimed that the scientist cannot say anything about how people should act; that only the humanist can tell us how to act. But we have not even mentioned the question of how people should act. What values, what standards, what criteria, should be used for choice? Many people at the present time are rejecting the elite documents and the classical canons. They are trying to make new canons and new ways of looking at things, trying to reject the past and to create the future. And we are faced, in our students, with a new kind of philistinism that we have not been able to see clearly because we approve of their concerns—their concerns for what one does about the ghetto, about poverty and misery throughout the world, about political corruption and political integrity. In the 1950's, the students were interested in how they were going to get rich. In both cases, their concern was and is with something practical, not with something theoretical; with something immediate, not with something timeless or lasting. If the humanists are going to do what they should do, they have somehow got to relate their documents to humane and lasting values— if you will, to moral values.

The question of the underlying deep structure, which I would take to mean in some sense universals—things that are "true," regardless of their expression—is one kind of question the humanist ought to ask. As we look at pop art, or the drama of the absurd, or the kind of art in life that the hippies live, one problem of the humanist is to relate these things for us to other human expressions. Peter Caws says that nobody reads Ovid or

Tacitus anymore, probably because the social context makes us feel that they no longer illustrate or meaningfully transmit to us their deep structures. But the deep structures are still there; it is really a question of vocabulary and the way you clothe the structures in order to illustrate them, to exemplify them.

It is clear that I have unwittingly touched a sensitive nerve. I avoided use of the phrase "eternal verities" because I do not believe there are any that we are in possession of or will be in possession of for a long time to come. I do not contend that the humanist should set up a standard of values to which all wise and honest men should adhere, although one sometimes gets the impression from reading humanistic writers that that is what they think they should do. I suggest that their task is to seek for values in the traditional manner of the scholar; to seek for underlying relationships, underlying homogeneity in the very diverse manifestations of human activity, including human artistic activity.

For example, in the case of what Mr. Caws called anti-art, I suspect that there is not any such thing as anti-art. If the human mind works only in certain ways, there are only certain things that human beings can do. I would plead that humanists emphasize their traditional concerns—their concerns with understanding, relating, and ordering values. It is not really true that you can just leave people to choose for themselves, unless they have as much understanding as possible. The more understanding any individual has, the more likely it is that he will choose a humane set of values. Cèline's works or Virginia Wolff's *Dalloway*—these points of view—need to be related in a scholarly and understanding way to *Lear* and *Hamlet*. It is not a question of teaching a single set of values. As far as I know, only a few persons in history have ever succeeded in propounding a set of values that was really helpful to other people. But it is possible for people to think about values, and I submit that it is the humanist's role to think about them and to try to be scholarly about them, as the biologist is a scholar about the workings of the body and of the brain.

MARTIN MALIA:
On the
Languages of
Humanistic
Studies

If we remain too abstract—and I must admit I am not very clear on how we can get to be more concrete—we will wind up functioning as amateur philosophers in a rather minor branch of philosophy. The concern with the methodology of the various kinds of academic disciplines is, historically speaking, a relatively recent phenomenon. I am not terribly informed on this development, but I have the impression that it began with the natural sciences, with the effort to figure out just what is the methodology of physics or biology, how they are similar, how they are different. What is the nature of the generalization that a scientist makes when he makes one?

The methodology of science is by now a well-developed discipline; very high-powered people have devoted their attention to it, and the descriptions or analyses of scientific explanation are now quite sophisticated. But this makes absolutely no difference in the way the natural scientist functions. Physicists and chemists and biologists do not have to know a thing about the philosophy of science to be first-rate physicists and chemists and biologists. If we are trying to get at the methodology or philosophy of the humanities in order to improve the way we do them on the basis of what goes on in the sciences and the social sciences, this minor philosophical inquiry will not be terribly useful. Let us take the example of the social sciences, and by this I mean disciplines like sociology, political science, anthropology, or history. Concern with the methodology of history is a relatively recent thing. There began to be a plethora of books on the subject starting some twenty or thirty years ago. At one time, I found this concern very intriguing and spent a fair amount of time on it. The central preoccupation very quickly became the concepts and categories of the historian: What is the nature of historical explanation? And after a few times around with this inquiry, it became clear that most of what was said was simply a minor branch of general epistemology and something more appropriate for a philosopher than for a historian to do. We were talking essentially about the logic of general discourse in these investigations, and, as in the natural sciences, it has not made any difference in the way history is written. The way history is done is changed by the injection of influences primarily from the social sciences. As I understand it, discussions about the methodology of the social sciences are now very much the rage. There has been considerable writing and a great amount of talk about "what is our method?" Here the origin of the concern seems to be that, to take sociology as an example, there has been a burgeoning of literature and theories in a short period of time—the last thirty or forty years—which has led to a sense of confusion about how sociologists work and what is the nature of what they are doing. So methodology was investigated, in part to clarify what they are doing, but also in the hope of eventually getting at a general method. Once one got this method, then all sorts of problems would automatically fall into some sort of general pattern, and one could start giving scientific answers rapidly. I think it is somewhat illusory to think in terms of getting this method, if I understand the quest rightly. What struck me about the nature of the conversations in question is that people very quickly began functioning as amateur philosophers, in effect investigating the logic of ordinary discourse.

I would be a little concerned if we got too far into an analysis of language; to me the languages of the humanities are the languages of ordinary discourse, the languages we talk every day. There is a big literature on this, going back to Plato and Aristotle, and we cannot compete. I hope that we can find some way to make things more concrete by restricting our attention to literary humanities or by finding some other problem that is less in the domain of general philosophy.

On the problem of defining the humanities, I should like to say that the definition we come up with will depend on the purpose we have. I do not think it would be terribly fruitful to try to devise in the abstract a definition of the humanities that would fit, for instance, the vast categories and subject matters that these disciplines encompass. In this connection, the sug-

MARTIN MALIA:
On the Languages of the Humanistic Studies

gestion that the humanities are concerned with elite documents or expressive documents would hold true for arts and letters, but not for something like history—if you want to call history a part of the humanities—because very often, and extremely often indeed in recent times, history is concerned essentially with very banal and trivial documents.

Historicism means the various things that people have historically taken it to mean, and it is quite simple to find out what it means to one person and what it means to another. The method, the language, the standards have meaning only in terms of a problem or a subject. They do not exist in the abstract. And to take that illustrious model that we are all consciously or unconsciously in awe of—the physical sciences—they did not start with a quest for *the* method or *the* right way to proceed. They wanted to know how stones fell or whether the sun went around the earth or the other way around. They started with a problem, and the method was devised in terms of the problem. The method was found to be good or bad in terms of whether it gave verifiable results of one sort or another. We have to define the subject or problem; namely, what are we talking about when we say the humanities?

I would like to suggest a rather crude definition, but one which might at least focus the problem. The humanities have a subject—the human condition, how people live, how they act, how they make decisions, what they consider good or bad, what they consider valuable or worthless; in other words, human affairs. This is talked about in terms of a medium—the medium is language in the sense of words as we use them in everyday life. Therefore in practice this would mean things like literature, history, philosophy, criticism. Let us not talk about the human condition in terms of forms, colors, or sounds—that is, in terms of the fine arts and music—not because they are not important or not humanistic, but because they add another dimension that makes things terribly complicated. Since we want to focus on the languages of the humanities, let us talk about things that are expressed in verbal form—literatures, philosophy, history, criticism.

Novels, poems, essays, criticisms, works of history, editorials in the newspapers—all use the language we speak, and the language we speak is a fluid and flexible thing. Words change over time. Nonetheless, all these words, no matter how slippery they are, reflect intelligible thought processes; they have meanings. They are not symbols; they are not like sounds; they are not like colors; they are in a category all their own.

I am convinced that the proper model for history and literary criticism, insofar as this is a historical subject, is the evolutionary one. I do not think, however, that the evolutionary model applies to all the different kinds of things that the historian does. It applies to what may be called the dialectic of the questions he asks. For example, Beard asked: Is the American constitution to be explained by the economic class interests of the people writing it?—a problem which, once it is set up, gets shot down, and so forth. The evolutionary model also applies to the changing awareness of what are relevant bodies of data. There is, however, a substratum of very fundamental kinds of judgment that does not change much at all. I will give just one example. Aristotle says somewhere that if liberty degenerates into anarchy, it ends in tyranny; this was, obviously, a long time ago. Even earlier, Thucydides has a famous section on the Corcyrean revolution which says roughly the same thing. Edmund Burke came along in 1789 or 1790 and wrote a book on the French Revolution, and one of its most essential themes is simply that if liberty degenerates into anarchy, it ends in tyranny. (He obviously says other things, too, about the organic development of society.) In the early-twentieth century, Crane Brinton came along, after having read Pareto and learned a lot about "residues" and "derivations"—sophisticated terms for talking about the irrational in human affairs—and wrote a book where he tried to analyze revolutions in terms of medical pathology. And when it is all over, one of the central

MARTIN MALIA: n the Languages the Humanistic Studies

themes that emerges is: When liberty degenerates into anarchy, it ends in tyranny.

This is what makes such history good—because, in a sort of *ad hoc,* empirical way, our experience with revolutionary upheavals indicates that is one of the most fundamental judgments we can make about them. What makes a good book of history, let us say about a previously poorly studied revolution (allowing for the fact that for the first fifty years after a major revolution most of the writing is largely polemical, soaked in the values of one or another party in the event), is when the author starts asking questions about the process, the mechanism according to which the event unfolded, and then begins to come up with something like this particular eternal verity I have talked about—or one such as de Tocqueville's dictum, apropos of the Old Regime in France, that the most dangerous time for a bad government is when it starts to reform itself (something that is rather illuminating about the United States in the sixties, I would say).

I agree with the statement that the subject matter of the humanities is the human experience. I tried to call it the human condition last night and was not allowed to get away with it—but now that the subject has been brought up again, I want to say I think this is quite true. The question remains: In terms of what do we study the human experience? The human experience obviously is everything—psychologists also study the human experience and in a way, I suppose, if you stretch it far enough, biologists or even chemists study the human experience, but biology and chemistry are not the humanities. There has to be an element of consciousness involved—the human consciousness. But then how do we distinguish the humanities from psychology or the social sciences? I would say the humanities study the conscious forms of the human experience in terms of what I previously tried to call the language of ordinary discourse—that is, everyday logic, sensibility, and judgment. And as a practical matter we do this through various kinds of cultural artifacts which we have been calling documents. Mr. Revelle got close to something very important in discussing the example of historical demography. This example illustrates very well the difference between studying the human experience in humanistic terms and studying it in scientific terms. Historical demography is not really history any more—it is the beginning of a new social science. It is the beginning of a new social science because it has taken a very limited problem, which is the way the natural sciences have proceeded since Galileo, and because this problem can be quantified in some approximate fashion. We can talk about it in terms of dependent and independent variables. Moreover, it has a practical use in that it can possibly shed light on demographic problems in modern underdeveloped countries. But the humanities—that is, the branch from which historical demography is now diverging: in this case, history—do not have a practical use of this sort. The use of the humanities, as Mr. Weil pointed out, is self-understanding. To put it another way, the only use of the humanities is education. The humanities give self-understanding, the understanding of others, the education of self, and the education of the next generation. They have no practical use in the sense of telling us where oil is or how the Egyptian population pattern might change in ten years.

To come back to this process of sciences branching off from the humanities, I think that historically speaking all the sciences have branched off from what were disciplines that we would now call humanistic. To take the most obvious example, the natural sciences: The natural sciences—first in the Greek world and then again in the medieval world—all branched off from philosophy. Originally there was no differentiation between physics and metaphysics—it was one great big grab bag, and it is well known that in the seventeenth century, when the physical sciences—which is the great paradigm that is haunting us all—got started, they were called natural philosophy. The natural sciences branched off precisely by focusing on questions that did not have

MARTIN MALIA: infinite implications; the focus was on manageable problems.

To take another end of this spectrum, the social sciences all branched off from history, beginning in the late-eighteenth century. The first one to branch off was economics, again by taking a relatively small manageable problem—how the market works. That is still pretty big, but it is not so big as the human experience, and eventually one can quantify it. Then later on sociology and political science broke away. Political science has been trying to "scientize" itself only in the past two or three decades, and the results are much less impressive.

Anthropology became a science by peeling off from a larger discipline—biology. And the process of peeling off in each case involved taking one relatively restricted aspect of the human experience in its social dimension —not in a psychological or aesthetic, but in a social dimension—so that fairly precise questions could be asked, questions that already anticipate the answer, and a describable method could be devised and put into a handbook and communicated to students. But the fact that the natural sciences branched off from philosophy and most of the social sciences from history does not mean that philosophy and history disappeared or became superfluous, and it does not mean they became "scientizable."

History is dealing with the human experience in its collective or social dimension. Philosophy is the most high-level, abstract, and "unscientizable" reflection on the human experience. These areas of concern always remain, and as knowledge gets more specialized and "scientizable"—if there is such a word—the humanities are seen as the great residual discipline where people continue to talk about the human experience in terms of the ordinary categories of logic, sensibility, and judgment, and this residual area will never go away; it will never get smaller. All sorts of things are added to it with time—it will never be suppressed, and I do not think its fundamental nature can change.

In addition, there exists a process of feedback from the newer to the older disciplines. After the social sciences branched off from history, they developed all sorts of systematic patterns about revolutions, economic crises, what have you, systematic patterns that are always immensely valuable in understanding specific cases. And history, as the residual in the area of the social dimension of the human experience, is concerned with specifics. The historian wants to know why the French Revolution happened, not the structure of revolutions in general. Thus the traditional rule-of-thumb historical judgments—the tropisms I mentioned earlier which are terribly useful—have been enormously enriched by the feedback of much more systematic patterns from the social sciences. Nonetheless, although the humanists who remain in the residual area of the human experience are usually made to feel inferior by these dramatic and spectacularly successful branch-offs, I do not think we should get too overwhelmed, because we are still dealing with the core of the whole thing.

I would hazard a guess that once a realm of humanistic study (or some peel-off from the humanities) begins to have a tangible usefulness, once it can affect, control, or move something one way or another in society, it begins to become something else; it is no longer a part of the humanities, but one of these terribly numerous offshoots. And it seems to me that the main use of the humanities is to give a kind of wisdom about what we are.

Some of the sciences and social sciences have separated themselves out much more successfully than others. Economics has, by virtue of its becoming more limited in the problems it undertakes. I would say that political science has done this least successfully, and I would put sociology somewhere in between. I was giving an overly abstract and perhaps excessively optimistic picture of the "scientization" of the social sciences, but something like this is going on, and certainly the people in the social sciences are quite consciously trying to proceed in this direction. That is one of the reasons why I think their high-level methodological discussion has preceded ours.

MARTIN MALIA: On the Languages of the Humanistic Studies

With respect to what I have called the residual area—history—new problems in the study of society come from two sources. First, they arise from the feedback from systematic social sciences. Marc Bloch has a radically new and exciting view of medieval feudalism because, in part, he read Durkheim. The second source of feedback is current events. Obviously when one reads about the population explosion in the newspaper every day, some historian is going to ask how the problem stood in the sixteenth, seventeenth, and eighteenth centuries—that is, in periods for which one can get data. The new problems and new advances in history do not come from something within the discipline itself. There are certain basic categories or modes of thought in the discipline that are as old as the Greeks. The same is true in aesthetics. I was rather gratified to hear Aristotle's name coming back so frequently this afternoon; that is the way it works in history—the new problems are generated by the peel-off of the disciplines and by current events.

I think the best case to show the way things peel off from the humanities is the most classical one: namely the physical sciences—how physics peeled off from philosophy. The weight of the literature on this subject is rather clear: Essentially, the way Galileo got started was by asking not the big question about the nature of the physical universe—what is its first cause and its final cause—but how those stones fall, something limited in scope, about which he could propose a clear answer that he could talk about in terms of independent and dependent variables.

I did not mean to suggest that the whole revolution took place in the lifetime of one man—Galileo. His work was obviously preceded by a long process of reflection within Scholastic philosophy that eventually set up the preconditions that made it possible in the seventeenth century to isolate a question much more completely than had been the case before.

What makes a good work of history as opposed to a bad one is the richness and verisimilitude with which the historian, functioning a little bit as an artist but also with a strong awareness of real social processes, makes the whole experience come alive as great social drama. One thing about the humanities is that we do not learn from experience. An individual does in the course of his life, but humanity in the aggregate does not, and that is why the humanists always have something new to say. Each generation must rediscover these eternal verities in its own context.

The Governance of
the Universities I

The first of these meetings was entirely without agenda. The conference started slowly. The tragic death of Martin Luther King was in the mind of everyone who spoke on that first day.

Participants
Daniel Bell
John Brademas
Jill Conway
Martin Duberman
Robert J. Glaser
Stephen R. Graubard
Hanna H. Gray
Carl Kaysen
Edward H. Levi
Martin Meyerson
Robert S. Morison
Talcott Parsons
Bruce L. Payne
David Riesman
Neil R. Rudenstine
Preble Stolz
David M. Wax

Historically universities have been part of an integrative system of society because they hold the values of society and pass them on. The universities have now become part of the adaptive system of society in terms of its innovation. Can they carry on both burdens? Can they be both integrative and adaptive? Logically, it seems to me, one would say that they cannot be both because the strains become too large in those terms. Should the whole research organization be re-examined so that something like an academy system, which the Russians have put forward, is the model? I am not making a judgment one way or the other; I am merely proposing that these are the kinds of questions to raise. Should there not be more and more government laboratories to take over the problems of science, rather than simply not-for-profit corporations which are attached to universities? You have a variety of models of which the university is only one: the academy system, government laboratories, university laboratories. All of these questions must be answered in reference to the framework of the purpose of a university.

To try to focus upon the function of the university and the new shape of the university, I would formulate the problems in this way. Since 1945, although one could date it earlier, a whole series of new functions has been thrust upon the university which raises the question: To whom is the university responsible? Practically without question, new functions were assumed by the university. It was almost taken for granted that the university would be the place for these tasks to be done. There was in part, I suspect, the sense of the tragedy of World War II and the continuing strain of the protracted conflict with the Communist world. There was also a sense of a mobilized posture. Because of a whole series of processes, including this mobilized posture, the university became expanded. You had the growth of various institutes, for example, to train people for Russian studies, Chinese studies—things which had never been considered as being intrinsically scholarly functions. But we needed Russian experts, Arabic specialists, and so forth. As part of the whole aura of science, the emphasis on research, you had a series of functions related to the design of defense—missile systems and questions of this kind. A great many of these functions were not just thrust upon the university, but were in some sense accepted as a social responsibility. By and large, there was a sense that scientists had to be involved in military problems and the science gap. M.I.T. became the center of a large political thrust against SAC. A huge bureaucratic struggle took place in the labyrinths of Washington in terms of setting up a

Distant Early Warning Line, a continental air defense, and various other measures. In large part, these programs were initiated by scientists using the university as a base. There was a great self-consciousness that the university would be an important political base in this regard. A whole series of things began to develop out of these elements. There was a feeling that mass higher education was needed. Professors considered themselves a new class and self-confidently accepted the argument that they were a new class.

Partly in jest, Mr. Kaysen asked whether, if the business corporation in a sense encompasses all the profit functions in society, the university encompasses all the not-for-profit functions. To some extent, this is quite truly happening. I thought it was quite interesting that Mr. Kaysen approves of this, although he runs a nineteenth-century institution. But Mr. Meyerson, who runs a twenty-first-century institution, retreats from this. Each is obviously searching for what he does not have in this respect.

In the last four or five years, there have been some very interesting reactions to the transformation the university has undergone during the last twenty-five years. The two major reactions have been in part because of the growth of a semiskilled intelligentsia who are, in a sense, the graduate students and the graduate assistants. They have suddenly discovered bureaucracy in the university. Bureaucracy is a dirty word, and to some extent it has been a bad situation for them, particularly in terms of the teaching conditions at the large state universities. Even in a place like Berkeley, something like 15 per cent of all classes are taught mostly by graduate assistants.

RIESMAN

What is so bad about that?

BELL

It is bad not so much because they are teaching, but because of the ambiguity of their status. By and large, they do not seem to have a feeling of being respected; they do not know exactly where they are going to go.

The university used to be a collegial institution, although it was never wholly that. It now becomes a bureaucratic institution, although it never wholly becomes that either. The second reaction has arisen in large measure because of the Vietnam war: the charge that the university has become part of the social system, part of the military-industrial complex in particular. If one looks at the figures, although they have been changing very radically, it turns out that more than half the research money comes from N.I.H., rather than the Defense Department. This reaction is symbolized by the campaign against the Institute of Defense Analysis and university ties to the military-industrial complex. It is quite clear in retrospect that I.D.A. was a very ambiguous thing. I.D.A. was set up by the Army to do weapons evaluations, and twelve universities were asked to lend their names to create a respectability for the undertaking. It is quite interesting that universities were asked to lend their names to a project in which they had no role. It was not as if they were in Brookhaven where they are really part of management or actively managing it the way they are, let's say, at Argonne. They were to provide window dressing.

One finds three different kinds of reactions about university activities; all are contradictory in a way and pose a series of problems. On the one hand, you get the extreme left-wing attitude that the university is corrupt, and you have got to destroy it because it is now a part of a rotten social system. This has become almost a universal reaction. Students toppled the city administration of Berlin after one boy was shot during demonstrations over the visit of the Shah of Iran. There were riots in about fifty German cities and a great sense of instability in the political system suddenly being introduced by the extraordinary tactic of student activism. The effect has not been so great in this country, but universities have been shaken—particularly Berkeley—by this kind of intense negativism, nihilism, reaction, romanticism, or idealism.

There is a curious second contradiction which is allied to the first and yet has an independent source: the plea for a new activism on the part of the university—namely, that the university get involved in race and urban affairs. There is a polarization between good activities, such as those related to urban affairs, and bad ones, related mostly to the university's ties to the military-industrial complex. Sometimes students and even faculty talk in terms of wanting the university to become tranquil, a place to teach; at the same time, they suddenly say that they want the university to undertake these good activities. Thus, you get the same kinds of pressures, but in terms of a good cause. To some extent the universities again, oddly enough, have not scrutinized the situation, but have gone along. Columbia is now setting up a large urban study center because Ford is putting up ten million dollars. There is nobody of real competence or interest in the university on this problem; we have gone outside now to bring in about ten people, all of whom have no relation to the university and many of whom are political figures. People at Columbia have suddenly decided that the university must be involved with Harlem. We have organized remedial classes and helped to organize small businesses.

The very same process which took place in the forties allied to defense and the Cold War and which created certain reactions has suddenly been brought forth again under the pressure of the urban situation and has been accepted without real forethought as to whether such activities are a part of the university's role.

The first reaction has been the one of destruction and nihilism; the second has been a new activism for good things; the third has been a mixed notion of retreat to teaching—to recapture somehow the old nineteenth-century ideal of the university. The old pastoral romance has now become a slum romance with the university as a great center of communion in which teacher and student come together. Here too, of course, you have the visions. You have those who talk about restoring the old student-teacher relationship and those who talk about destroying it so that the university becomes a place where everybody goes around touching and feeling one another and expressing joy. The latter becomes a form of group therapy with the implicit notion that a teacher is not a competent person who knows more than a student and enters into an adult relationship with him. The assumption is that one cannot learn from another person; one can only learn together.

If one wants to define the purpose of the university, one must reach this definition by deciding whether the university should accept a whole series of activities which have been thrust upon it simply because it has been the one institution available to carry on these burdens.

DUBERMAN

I am surprised at Mr. Bell's terminology and at some of his specific examples. Some of the things that he labeled "romantic" would from my point of view require more definition. I am not sure that this is one of the two main camps on any given issue.

KAYSEN

I wanted to comment on two of the things that Mr. Bell said. I do not differ with his description at all; I do think, however, that some details of this picture are worth mentioning. First of all, I see the origin of all this in the World War II experience, especially that of two, perhaps three, institutions. The two most striking ones are the Manhattan Project and the M.I.T. Radiation Lab which did radar work. There were also some smaller projects and lots of committees—the National Defense Research Committee, for example. Perhaps the most significant aspect of the Manhattan Project and the Radiation Lab is that they were probably the first serious interdisciplinary efforts on a large scale. They put together people who normally did not work together in the ordinary structure of the university. Bell Labs achieved this only to a moderate extent, and it was departmentalized. Bell Labs changed its structure a lot after the war, in part by learning from this kind of experience. The people involved were not dragooned; they were eager to do it for two different kinds of reasons. One, of course, was political. We have to think back not to 1945, but to 1939, to the Nazis and the war in Europe, and to the significant element of refugee input into American science, which had a lot to do with the spirit of many of these enterprises. Einstein, Leo Szilard, and many others came here after they had been driven out of Europe. The problems they worked on were intellectually difficult. The natural scientists and later the social scientists discovered for the first time that things which ordinarily would have earned their contempt as engineering, as applied science, were technically and intellectually quite interesting. George Kistiakowsky, a very good abstract physical chemist, found that it was fun to be an engineer and to design a particular device—namely, the first nuclear bomb—and to show that the theoretical physical chemist was a better engineer than all the engineers.

RIESMAN

European baseball.

KAYSEN

Yes, that is right. There were other organizations in which this kind of thing went on. One of them was the O.S.S. which had many qualities of an interdisciplinary character. It was the first interdisciplinary social science venture; this was Harvard's march on Washington. Aside from General Donovan, what you had was a group of Harvard professors in history, in economics, in political science, in certain languages who were put together to create the country's first intelligence analysis organization. This amateur organization—and this may be a parochial view—was much better than the professional civil service organizations. Just as George Kistiakowsky and Ed Purcell and Isadore Rabi turned out to be much better electrical engineers, mechanical engineers, chemical engineers, factory managers, than the engineers and the factory managers, professors of French and Renaissance history turned out to be much better intelligence agents than people in the intelligence profession.

Another organization important to this history is RAND.

RAND was set up outside a university context, but it had this same approach—interdisciplinary problem-solving. Its first administrators were quite bright and followed a basic academic rule. They said that RAND should not care about the problems it was going to solve, but should create a situation so attractive that lots of bright academics come, regardless of what they do. Thus, Tjalling Koopmans, professor of economics at Yale, a very theoretical person and, incidentally, a pacifist, went to RAND and wrote theoretical papers on the general characteristics of economic equilibrium systems, and RAND was smart enough to pay him to do that. And because Koopmans was there, young men came there and got involved in more concrete projects, so the institution became inventive.

If you look at the university during the postwar period for comparison, you see a substantial shift in the disciplinary composition of the status order and numbers in a faculty. In the twenties, the professors of languages, literature, and history would have been the people who were heads of faculty committees at any of the great universities. Now the natural and social scientists carry the heavy weight. Part of this is statistical change; these are the departments that have grown the most. The social scientists are drawn to the kind of work we have been talking about because, in a certain sense, it is field work for them. If you get involved in a city's problems and are an urban sociologist, you have access to data that you would not have in the university context. You actually generate data which you would have a hard time generating in an academic setting. Many natural scientists cannot do science full time; it is too intensive, too intellectually demanding. So much applied activity is recreation; it is baseball, as Mr. Riesman remarked. One of my colleagues who is a most theoretically oriented man, Freeman Dyson, designs rockets because he enjoys it, although he is really interested in quantum electrodynamics.

The phenomenon to which Mr. Bell rightly calls attention represents not only society pushing functions on the university, but the university reaching out to repeat experiences which its active constituents—the faculty—enjoy. If one looked carefully into the history of most of these enterprises—and this is a conjecture—a faculty member, rather than a dean or an administrator, would appear as the active entrepreneur, as the person who has made the bridge between the foundation, the government department, the corporation, and so on. The Carnegie Corporation and the State Department thought that it was a good idea to have a Russian research center, but without the entrepreneurship of Clyde Kluckhohn we would not have had one.

I would round off these perhaps rambling observations by saying that this activity will be bounded in some dimensions; unless you find some correspondence between the natural intellectual interests of some group of faculty members and what they feel professional competence is, it will tend not to take root. On the other hand, that may be small consolation, because what we define as the legitimate intellectual interests of professors has widened enormously. There is a general belief in the society that there is nothing to which intellectual expertise is not relevant.

PARSONS

May I add a footnote? The Russian Research Center at Harvard and the Russian Institute at Columbia were established at the beginning of the Cold War. The people who came into them had a dual orientation with respect to public service. They thought knowledge of Soviet society would be extremely useful in case of serious trouble, but were, on the other hand, disposed to hope that institutes of this kind could help in achieving a *rapprochement* and an easing of the conflict. In other words, these were not warmongers who, having knocked the Nazis out, decided they were going after the Russians. I doubt very much if people on the left are willing to give the professoriate that kind of credit. Some of us have been in contact through Pugwash and other channels with a major effort of American natural scientists to work with the Russians on disarmament problems and to go quite far in that connection.

MEYERSON

Mr. Bell mentioned that his university had accepted a large grant of money to establish an agenda for the university, an agenda which did not derive from the interests of the members of the university. When we talk about the ethic of the university and the governance of the university, are we talking of the university as an environment that enables a great many intellectual activities to take place, or are we talking about a hierarchical organization in which there are central decisions on how resources get allocated and to what purposes? Mr. Levi said earlier that, in a sense, the best university is the least governed one. If that is the operating principle, there would be little opportunity to make these central decisions; the decisions would essentially have to be made on a decentralized basis. Are we talking about different kinds of decisions? Are there decisions that can appropriately be determined centrally, while others ought only to arise from the members of the institutions? And if we talk about membership, are we talking only about the current membership, or about a future membership as well?

BELL

A university plays a crucial role in society not only because of the problems of service that are all evident and real, but for a latent reason. One of the aspects of any modern society is the breakdown of any notion of ritual, a breakdown of a place—call it transcendental place, call it symbolic place—which is meaningful to people and provides some sense of something of importance that goes beyond the immediate self-interest and the notion of the person *per se*. Increasingly one finds, particularly among middle-class people, the notion that there are no places which mark out the passages in one's life. We all talk about the breakdown of religion and about the search for religious things. For most people there are no distinctive places any more; patriotism does not provide you with a sense of identification; religious rites no longer give you that sense of anchorage. To some extent, for a lot of people the university has become, without this being explicated, the transcendental institution in society because it seems to promise the notion of a community. It is a place in which people feel an attachment to something beyond themselves—scholarship, learning, books, ideas, the past. The university has some sense of reverence attached to it; it is a place where you have colleagues and engage in activities which are satisfying to you in a very emotional way. A lot of student agitation about the university comes out of the fear that they want the university to be such a place, but feel that somehow it is not living up to their expectation. In terms of a long-run structural feature of the society, there is a sense in which a long time ago people thought that the corporation would perhaps become a community, an occupational community replacing the family. This idea broke down quickly because the corporation was not that kind of institution. The university, however, does have that kind of resonance as a community, but this raises a real question—namely, what are the limits of a community in this sense? The university cannot be a total community because it cannot be the whole of one's life, but it has to be a large part of one's life.

When the university becomes more and more a center for culture in society—not just for science or for social science—theaters become part of the university, as do the little magazines. It is accepted somehow that the university will take such things under its wing. More importantly, as universities grow, they should provide places wherein people can spend a large part of their lives. Lectures, theaters, music supply a sense of attachment.

If one asked what the university is for, I would say it is for more than scholarship, although it is for that; it is for more than teaching, although it is for that. It is basically for a sense of a community. One of the great attractions of the University of Chicago has always been its sense of being a community. It provides a place where people live near one another without necessarily becoming conformists, without necessarily becoming small-towners. It engenders a sense of attachment which makes you feel that a large part of your life is being lived in a meaningful way. All of us still hold to a certain extent the Western ideal that work and home and place ought to have some sense of unity; that somehow the sundering of work and place, and work and family, and work and emotional attachment is wrong. The university is the one place that has been able to recapture the feeling that work and place and home and colleagues and friendships do have this kind of unity. If the university does provide such a sense, you have certain baselines for identifying what you want to take out of the university and what does not belong in a university atmosphere.

RIESMAN

When Mr. Parsons and I were involved in filling a chair in social ethics at Harvard, we found it very difficult to find nonschizophrenic, disciplined, cogent academicians who connected their work and their emotional concerns in a way that was not metaphorically simplistic or methodologically trivial. When I consider this in light of our discussion, it seems to me that the ethical, human, and integrative problems look different in different fields of knowledge. I cannot imagine most social scientists being so brilliant as most natural scientists, getting their work done and then wanting to play around. Social scientists and certainly most people in the humanities are just too dumb; moreover, their work is too involving, and they are too narcissistic about it. I have perhaps an idyllic image of the natural scientists as being more sanguine, more team-prepared, less concerned with small status struggles. Theoretical physicists have more prestige than metallurgists, but still do not hesitate to associate with them and indeed they enjoy these opportunities. As Messrs. Meyerson, Kaysen, and Bell have suggested, the social scientists enjoy downward social mobility—in other words, slumming. They are perfectly willing to go into Harlem or the urban center; that is moral or that is legitimate.

My experience in teaching, both at Chicago and Harvard, suggests that it is almost impossible to get students not to consider themselves corrupted if they go upward. If I tell a student to go and work in the Federal Government to see how problems of disarmament are really discussed, I meet the feeling that that is corrupt. When I say he has got to know the language in which these problems are discussed in the government if he is to have any impact, I encounter the same pastoral romanticism to which Mr. Bell referred.

This makes for enormous disparities among the different academic guilds. How does one's affective life get touched in one's work? What is the impact of that attachment on the relations to one's colleagues and to general society? What contacts are regarded as field work or educative? The answers that the natural sciences do not give, but that many of the concerned people in the social sciences and humanities do give are precisely illustrated by the Bell-Duberman exchange. In the humanities and the social sciences, the growth of "total individuals" is regarded as a perfectly legitimate, if dangerous, academic concern. In the natural sciences, bringing along colleagues who are par for brilliance is, on the whole, the only kind of teaching which has legitimacy. One of the problems of our discussion reflects the disparities in intelligence, ease of playfulness, locales of preoccupation as these change and as these reflect the way people get differentially drawn into different fields.

GRAUBARD

Messrs. Bell, Riesman, and Kaysen all see roles for the faculty in the university which they have been able to find for themselves or have been asked to assume. What is really interesting in these accounts is that there has been no such opportunity for the student. Despite the change in the character of the institution and the community in which the student is temporarily a member, it was assumed that he would be willing to go on with somewhat less attention than he had received simply because his professors had so many other competing interests. It is the students, however, who are saying in effect that they now want to become more active. In short, the student no longer accepts the idea of the university as a moratorium; he resists the notion that his university experience should be a waiting period during which time he acquires certain capabilities. At a meeting *Dædalus* recently had, very able graduate students, who could go on doing "their thing" the way that graduate students did in the 1950's, expressed a firm resolve no longer to do so. They felt the obligation, the need, to get involved, but the university has not been able to resolve the question of how to involve them, nor is it at all clear how the rest of the university community will react to their involvement or what the university will be like when they are involved.

BELL

It seems to me that two different things are happening here. Students resent faculty, and yet faculty help them; both are valid reactions. The important question is whether the student is searching for an identity or for an education. The two are very different things. In many cases, involvement in activism, which is poignant and real motivationally, is more a search for an identity than for education. Education is a confrontation not so much with events as with a teacher and a tradition. You have no way of holding or judging events. Although you do get a sense of emotional satisfaction that relates to these events, I am not sure you educate yourself through them. You would have the ideal process if a highly motivated and activist student followed these inclinations and afterward had a chance to spend two years as a Rhodes Scholar or as a fellow. He would then have had a sense of experience, of finding himself. When he came back and confronted a tradition, he could relate himself to it this way. I am not saying that the two are necessarily antagonistic, but in most cases you do not have the opportunity to do both. Then the question is what do you want to do. Are you asking a student to say education is a tradition; that it is a coherent notion of a rationally defensible body of learning to which he has to relate himself? Or are you asking him to find himself in terms of some antipodes? It may well be that something is wrong with the whole university system if it cannot trust itself.

KAYSEN

Obviously the more advanced graduate students, those writing dissertations, become apprentices within the system. There is, however, an important qualification here. In the natural sciences and to some extent in the harder social sciences, professors are allowed to play with applied problems, but their graduate students are not; they must do their proper, methodologically correct academic business. With that exception, the more advanced and better graduate students are allowed to participate in the academic process. The undergraduate is not primarily because that would be too expensive; it would take too much of the professor's time and so on. There are exceptions to this. In the bursary system at Yale, a student has to give a certain amount of time working as a research assistant to a professor in return for a scholarship. This is the way it is formally defined; in fact, of course, it is a great opportunity, and the bursary boy obviously gets a much better educational experience than he would if he did not have this opportunity. On the whole, a Harvard professor who tutors students treats them a little at arm's length, while at Yale the same man with the same student might involve that boy in his own work and therefore teach him in a much more serious and effective way. These are, of course, terrifically expensive educational devices which cannot be widely used. Thus, you return to the question of the economics of this whole process, to the cost structure of such educational arrangements, and to the scale at which it might be possible to provide them.

PAYNE

One of the things that ought to be noted, and has not been, is that while students are at an institution, they are very often more involved with and more committed to it than is the faculty. The faculty is off on the kinds of projects that have been talked about, it is off to Washington, it has its own methods, and so forth. The difference between the search for identity or for education is not the real problem. If students at a university are offered a chance to get an education, if they find a teacher, most of them will discover their identity there. Most of the time the students that I run into, even at Yale, do not find teachers. I agree with what Mr. Kaysen says about the bursary system; it can work well. Some tutorial systems have also worked well. Nevertheless, not many people have that kind of experience. If you are going to confront the teacher and the tradition, the teacher has got to be there—at least through intermediaries. I think that graduate-student teaching makes sense if it is part of a cooperative enterprise in which some of the best minds on the faculty take part. I went into political science because I found two honest men at Berkeley in my sophomore year who both happened to be in the political science department. I was able to confront them not in a situation divorced from student activism, but in an educational process. I find very few models for that kind of teaching. I do find an academic culture in which people are involved in a great many other things and have a great many interests beyond the boundaries of the campuses, interests generally opposed to teaching. It distresses me that at Yale, where you have got these people around, the teaching does not happen because there is a reserve, a backing off from it. The possibilities for this kind of teaching exist on most campuses, but are seldom used. There are not many people in academic positions who are concerned enough to teach in this way.

DUBERMAN

There are, perhaps, some basic disagreements here about what education is and, therefore, about the function of the university. I am not clear, for example, as to what Mr. Bell meant by distinguishing education and identity. In terms of what he did say about education, I am not convinced that his definitions are mine. He defined education as the student's confrontation with a teacher and a tradition, or, as he alternately put it, the way in which a student relates himself to a rational body of learning. It seems to me that this is one part of education, but I would prefer to define education more broadly as the student's confrontation with himself. This can be aided by a confrontation with a rational body of learning or tradition, but the latter, taken alone, is not all there is to the process of self-confrontation. Indeed, I am increasingly convinced that the content of university work has little to do with self-confrontation. Other things that go on either in the classroom or outside it contribute more to self-awareness and individual growth than the actual discussion and transmission of information.

PARSONS

Does this mean that the intellectual content is completely dispensable?

DUBERMAN

No.

PARSONS

This is a crucial point. It is a very defensible position to hold that confrontation with a teacher and a tradition is not all there is to education. Higher education without intellectual content, however, has never made great cultural history, just as religion without a theological tradition has never made great cultural history.

DUBERMAN

If the function of the university should be the accumulation and transmission of knowledge, and if a great deal of emphasis is put on that function, the students' needs must be and traditionally have been almost entirely ignored.

PARSONS

I do not think either Mr. Bell or I said that.

BELL

I think there are differences, and I would like to explore the lines further. If you are talking about education, I would emphasize the intellectual content as being primary. This does not mean that there is only one body of thought or one tradition, but at some point you have to put your thoughts together in a coherent way that is rationally defensible. I also said, however, that this confrontation can only take place within a community, a community which makes the acquisition of a body of knowledge a satisfactory process. I do find a little distressing the tendency to consider feeling and emotion and the notion of individual growth as an end, because the phrase "individual growth" becomes intransigent. You live for something; you do not relate only to persons. A teacher is a model who embodies something; he is not just an abstract person who is satisfactory in terms of his learning, his style, the way he has mastered a body of thought. He has lived a satisfactory life in his way. The old word *magister*, or master, carries with it an important concept; you master *something*. To use a Jamesian idea, you do not have a state of consciousness; you have a consciousness of something. You have a consciousness of an intellectual tradition—with all the emotional content that learning or recognizing this implies. This may be a very personal statement since I grew up in a highly Talmudic way of life in which you mastered with your thumb, one way or the other. You get a large thumb this way; you may not get a green thumb, but you may get an enlarged calloused one from turning book pages. That may be a special kind of bias, and yet it is one that is central to the history of Western culture. You cannot throw out the whole Western culture in the unthinking way that is now being done. You must have standards to make any kind of judgment, and thus you must have something that is rationally defensible as to what a standard is.

LEVI

Very strange things have happened over the years. We used to put facts against theories. Theories were good; facts were bad. Now theories have become facts in the lexicon, because they stand for the rational. Thus there is a kind of disrespect for the theoretical structure, which in itself mirrors the problem that we were talking about before—namely, the scope of the university. It is a great mistake to say that one of the jobs of the university is to provide a place where people from eighteen to twenty-five are to be kept while they grow up. That never was regarded as the best way to put it, and yet that is often suggested today. Another mistake we have made is to overemphasize the academic role as the sole way of learning. Moreover, the notion that the university should pre-empt all of the student's time in the sense of community or whatever is also a great mistake and contains elements of snobbism. One would hope that a sense of community could be developed without imposing all the operations upon the university. One of the tragedies is that universities exist in communities that do not themselves stimulate the kinds of operations which would be desirable. These operations are, therefore, put into the universities, and as soon as you do that, you get an extraordinarily peculiar kind of tension.

One of our problems is to cut down the size of the university at the same time that we increase the importance of the community. We must free the student from the notion that the university is the only path, or from the idea that when he elects to follow this path, it must be his total preoccupation. I do not want to belittle the problems of teaching. This is a constant struggle. There is, however, little to show that teaching and education are more difficult or less successful now than they used to be. I am not sure that education should come easily to anybody. The report that one or two or three faculty strive for stimulation is, I think, quite natural. Books can do that. Other things can do it as part of growing up.

As we talk about these things, the cost element should be regarded—not for the sake of universities, but for the sake of the society. From an educational standpoint, the tradition of large first- and second-year classes, particularly in the state institutions, and then more individual attention in the third and fourth years is absolutely wrong. If you had all the resources that you wished, you ought to be able to turn it the other way around. You cannot look at this question only in terms of universities, because the great problems of this society deal with primary and secondary education, and we should not forget that. The universities should not be hogs about educational resources. There is an enormous selfishness in pouring money into universities where it is not essential.

MORISON

I just have a couple of footnotes on the two-culture problem. There are some real differences between scientists and people in other parts of the university. I do think Mr. Riesman is right in saying that natural scientists have more time to go down to Washington and to engage in social things, but it may not necessarily be because they are brighter. It may have to do with the structure of science. In science a person can be assured of a long career by having done one brilliant piece of work. In physics, they very frequently do this when they are twenty-eight or so and then get a Nobel Prize when they are thirty-two. They feel satisfied at having done one or two or three things. Historians, on the other hand, take on the job, say, of understanding the nineteenth century, and there is no end to it. The structural differences are also important. A physicist or even a biologist has a lab with several people in it, and so he can reach the stage fairly early in life when his principal job is to talk with these people and get things set up, and then go to Washington. Thus he leads two different lives. Also, he has become rather deliberate in separating the second life from the first life, partly because of our worries about having classified projects on the campus. We deliberately separate graduate students from the secret things we engage in because we think it is bad education to have them do secret things. Many universities have now said that they are not going to have any classified research on the campus. Faculty members who want to do that go out and consult with industry, or work with RAND, or do something of that sort, and then come back to the university. This does not separate them from their students, although it does separate students from secret work.

When the scientist is on campus, he may have a fairly intimate relationship with his students, because he can bring them into his lab in a way that the professor of history does not bring them into his carrel. In science, the tutorial relationship is almost automatic, at least in the junior and senior years with the good student. The teacher welcomes that relationship because the student actually produces something for him. Consequently, students in the sciences begin to identify themselves as scientists relatively early and feel fairly satisfied. As far as I am able to observe in my institution, science students do not become activists in the political sense. The number of biologists who are picketing is extremely small. I worry, however, that they may become prematurely mature, so to speak. They never confront themselves as political people or as scientists or as emotional human beings. They get their emotional satisfaction out of identifying themselves with their profession, and I do not know whether that is good or bad. I do not know whether we are really educating our people or not. They seem to be happy, and they get happier as they go through college. By the time they are seniors, they feel great, but I am not sure how much they are going to contribute to society.

MEYERSON

In talking about the distinction between the search for identity and the search for education, Mr. Bell was urging both in his nostalgic approach to the University of Chicago. At least it was able to provide both ventures for faculty and probably for very many of the students there.

There is a conventional wisdom today which suggests that the private universities, with the exception of a handful, will cease to be: that Boston University probably cannot survive; that Northwestern will have the greatest difficulties; that Washington University in St. Louis will probably move in the same path of economic demise; that despite various efforts to provide some public compensation to students or to institutions, there will not be enough funds to enable the private institutions—apart from the Harvard's, the Yale's, the Stanford's, the Columbia's, the Chicago's —to exist in the future. As a result, in this view, there will have to be more and more public higher education, which will almost inevitably carry with it certain kinds of controls.

The concern about controls is relevant to a point Mr. Bell made: Can the university be both integrative and adaptive? Can a university that is to be a transmitter of the social values of a culture also be an innovative agent for radical change in the state of that culture? Again and again we see that there are great constraints upon such innovation where there are public controls. These same constraints also exist where there is private control through donors and other means, although they are more marked in the public sphere than in the private. These constraints run counter to the tradition of the opportunities for an innovative social role for universities, which is one of the reasons why I think that for many kinds of innovation we would be well-advised to look elsewhere. In the case of universities, we are dealing with institutions that are very fragile. Throughout history, they have been the transmitters of the existing social order—whether it be a religious social order or the kind of landed gentry aristocracy that the British universities had maintained. I would question whether the university can both obtain the level of resources that the future will require and be socially innovative. It can be educationally innovative in areas that people do not care about, but can it be the instrument of radical social change?

CONWAY

People have spoken of education as confrontation with a body of learning and as a process of self-confrontation. It has been generally accepted that the essential factor in achieving both confrontations is the teacher as model. I would like to point out that a large part of the student constituency of state universities and many private ones, too, is feminine, and that these students do not have many women teachers as models. This is an important educational problem, one which leads at present to a misdirection or inefficient use of educational resources. The amount of money that is invested in educating women today on the undergraduate and graduate level does not produce the kind of service to society that it could. One reason for this is that there are very few models in contemporary society for educated women to follow.

RUDENSTINE

The idea that the teacher is a model is probably a new one. The teacher used to be rather despised. Certainly in the twenties and thirties, the last person a great many people on campus would have looked to for their model would have been a teacher. Now the teacher is a professional model and an idealistic model in a way he has not been before. That role suggests an inordinate burden on the teacher and the university, and we are suffering from the strains of it.

Mr. Levi's idea of demythologizing the university and the teacher's role is a healthy one, but how we go about doing that is another problem. Part of the reason for this is that the career options open to students are no longer so secure, so acceptable, or so predictable as they once were. In the thirties somehow you could take your education for granted because you could take your career for granted. You did not have to fight every inch of the way with your teachers or their courses. You did not make those kinds of demands, because somehow education was what it always was and then you went off to your career. Now most undergraduates are puzzled about what they should do. The government is unacceptable; business is unacceptable. This is to speak hyperbolically, but most of the "ordinary" careers are no longer the place to go. People think much more now in terms of high-school teaching, inner-city projects, the Peace Corps, or Vista. Careers are in such flux that there are no clear options as to what would integrate one's ideals and one's work. Somehow the burden of integrating ideals and work got transferred to the university and university careers during the fifties. Students are still looking to the universities, but they are now beginning to be disillusioned. They are discovering that the teachers are not all they thought they were, that the universities are not all they thought they were. It would be good to demythologize both the teacher and the university. It would also be good if there were other career options so that this enormous burden would not be placed on a university.

Governance of the Universities I

BELL

When ritual and the order of your life break down, you look for some definition of reality and a confirmation by significant others. People need some sense of confirmation in this respect; somebody confirms you as to what you are in terms of whatever statuses you achieve. This is why people turn to the university. Lacking the various other kinds of persons who would be the significant others to confirm them, the teacher becomes the model. Mr. Rudenstine wants to demythologize this role; I, however, think it is a valid role simply because a university degree does become a valid model of an acceptable life.

RUDENSTINE

If you are an academic.

BELL

No. For a period of time, most of the young people in this country and certainly the brightest ones are going to be spending a significant part of their time in the academic world. It is important for that very reason not to demythologize the professor. That does not mean that you put him on a pedestal or anything of this sort. You need an aesthetic distance in order to be yourself and not be swamped by students. In order to be a confirming other for them, you must maintain this distance. To that extent, I go against Mr. Levi's plea that this role be restricted. I think one should enlarge it because there is perhaps no place other than the university in contemporary society where a confirming other can be found.

KAYSEN

In taking up Mr. Duberman's point, I have to ask myself the old question: Compared with what? He has in mind presumably some alternative ways of growing up, some alternative places to be between eighteen and twenty-five. In an earlier stage of social organization, growing up usually took place in an apprenticeship context where one combined the learning of some body of knowledge—not necessarily a traditional body of knowledge, but sometimes a body of knowledge with considerable abstract content—with the mastering of nonintellectual techniques: physical skills, interpersonal skills, whatnot.

A politician in the seventeenth or eighteenth century came to be a politician by being drawn from a certain class—in most polities, the aristocracy. He went to Court as a page or had some relation to the political establishment; in some context he would read and associate with learned men; he would also learn all kinds of interpersonal skills and some technical military skills, because every politician was expected to have some share of them. He did this in a way which was mostly highly concrete. There was little reflective discussion about political life, although there was some. Today you learn to be a politician in the university.

More and more skills have become intellectual skills. Again, let me draw some samples from science. A fellow named Fieser at Harvard used to dazzle students by doing an organic analysis very quickly using a lot of brilliant manual techniques. Now there is a spectrometer that does the whole thing more or less automatically. Thus, rather complicated manual techniques have been replaced by purely intellectual ones.

A particular element of the apprenticeship relation has to do with the physical handling of things in a community enterprise. I remember when the first synchrotron was built at Tech. It was a pretty sloppy thing; you would go down there, and somebody would be tinkering with it with a screwdriver. You cannot do that now; you have to call in the engineering department. The university, as well as the general society, has increasingly substituted intellectual techniques and bodies of knowledge for these other techniques. What is going on in the university is also going on in the society.

Most of the jobs that most college graduates will undertake are undergoing this transformation. That means that the university and academic work provide two kinds of role models and techniques: intellectual techniques and interpersonal techniques. If you accept this projection, although it is clearly exaggerated, the existentialist anxiety that we are being provided a role model only for a tiny fraction of the world which goes into the academic profession misses the point of what is going on.

Business is going to absorb for some time to come the largest number of people who get degrees from colleges. These graduates will go into jobs where they must be able to absorb a certain amount of information and translate it into a decision. They will operate in a context in which they must have some skill in interpersonal relations. All that they learn at the university, whatever they study.

DUBERMAN

I might be making reference to a rather narrow group of students—those I have seen in the humanities and the social sciences.

It may well be that students in the natural, physical, or biological sciences do not react in the same way. Nevertheless, the students I have known do not feel that their four years in college are a growth experience, nor do they feel that they are a learning experience. I could not agree more with Mr. Bell that growth and learning come from confrontation with significant others; most of these students, however, feel, as Mr. Payne has already said, that they do not come into contact with significant others on the faculty. What they come in contact with instead are specialists in academic learning; they absorb a body of knowledge about a particular field. Often in the humanities and the social sciences, that body of information does not appear to them to be noticeably relevant either to their needs or to those of society.

I am not at all clear about what can be done to correct this situation, but from my experience I am sure that it poses a real problem. From the point of view of most graduating seniors, the university has not been the best place for them to have spent the preceding three or four years. They do not feel that the university structure has met their needs. They often would agree that they had been educated in terms of Mr. Bell's definition. They have been put in touch with a rational body of learning or a tradition, but they doubt the relevance of the tradition; they doubt how relevant or how rational the body of learning is; and they also doubt whether they have been put in touch with whole people. They see their professors only for an occasional hour or so; they see them dealing with specific matters in terms of the mastery of a particular subject. They do not see, as they used to in certain progressive colleges like Antioch or Black Mountain, whole people performing in a large variety of areas. It seems to me that until they see such people and such variety, they will not have had confrontation with significant others.

PARSONS

I should like to come back to Mr. Meyerson's remarks. The question of whether the university can or cannot be innovative is an important one. I am speaking, however, more in terms of the university's cultural and social impact than of its effect on the individual student, which is a somewhat different problem. The greater need for public support and the general differentiation of the academic world from other sectors of society clearly mean that the university is developing a new combination of independence and dependence. I myself think that the university's structural autonomy is exceedingly important. If it is specialized in its traditional functions, teaching and advancement of knowledge, it cannot in itself be politically powerful. It cannot itself produce wealth on a big scale. It inevitably becomes dependent on non-academic agencies for its position in the society. Cases where this has not been true are largely those where higher education has been closely bound up with some special elite—as was true, for example, of Oxford and Cambridge in the late-nineteenth and early-twentieth centuries.

In this situation, only in generally revolutionary eras can the academic world be a spearhead of what is often referred to as fundamental social change directed, explicitly and in the short run, against the dominant institutions of the society. Over a century or two a university may deeply undermine the authority system of an earlier time, but it cannot directly challenge it and overthrow it. This is not the nature of its function. If it is to be innovative, it has to change the cultural base of the society.

We have to a degree institutionalized an important balance between constancy of pattern and innovation. The prototype is the advancement of knowledge, but carried on within a relatively stable, though changing, institutional framework. Beyond that there is obviously the dissemination of the implications of these changes through the education of people in a variety of ways. The distinction between being innovative in the sense of being essentially a political source for relatively immediate change and being innovative through these more indirect channels is extremely important. Doing something about the urban ghetto is an intermediate thing, but I quite agree with Mr. Meyerson that universities and students will not be the primary agents in the mitigation of ghetto and poverty problems. They can be an important stimulus, symbolically showing the way, but if there is to be a big push, it will probably have to be governmental.

RIESMAN

I want to pick up the financial and demographic problems on which Mr. Levi touched, and then try to link both to the demanding ideal of academic responsiveness to constituencies. I keep asking state university people: "How many on your faculty earn more than the governor? And how long can this go on?" In the case of the private universities, faculty are only beginning to get more than the trustees. How long can that go on? There are only a few places where it can go on at all. Faculty are more and more in demand because of the social needs we have been talking about. The resultant pressure from students is likely to create a faculty power backlash which will seldom manifest itself in formal counterattack, certainly not in support of Reaganism, but very likely in personal flight from demanding to less demanding settings. Yale, Princeton, Stanford—places like that can afford the luxury of having some concerned faculty because by and large those people do not want to go anywhere else. I do see a potential flight of faculty from places like Antioch because they share the Duberman ethos in some degree—the feeling that the students' claims are just. I know I feel this. The model I see for myself is that of racial integration—that is, student integration has some of the same problems of tipping the neighborhood. Negroes spread from the Black Belt of Chicago southward to the University of Chicago because this is liberal, nonviolent terrain; they do not go to Cicero or westward against the Polish or the Italian communities. Similarly, the vulnerable faculty go to the place where they will be cannibalized and not to the slightly more guarded gates where less is expected of them vis-à-vis demanding students.

On the one hand, universities seek faculty to do less and less for more and more; on the other, students ask faculty to do more and more for fewer elite students. This leaves the great bulk of institutions, many of whose faculties come out of the ethic of wanting to teach high school. Although I find many more of my students wanting to teach at Tugaloo or other Negro colleges than in a commuter urban poor-white college with some Negroes, nevertheless there is some effort to character these institutions with the ethic of total involvement and commitment. I do not have an answer to the issues I raise. How does one put together what I, along with Mr. Levi, believe to be a declining luminescence of the university with the increase in what is being asked of that fraction of academic men who have been socialized to want to teach? Such men find it hard not to respond, and yet (as in the case of racial integration) they feel that they are paying a personal price for previous cultural neglect.

PAYNE

It is clear from Mr. Kaysen's remarks that skills are imparted; it is also clear that we are in trouble when we risk meeting students even less than we are meeting them now, when we risk professors and students doing less than they are doing now about things they care about. How good is the educational community? What determines its quality? I am opposed to demythologizing the university. I want to remythologize it—professors, presidents, and students. There is nothing wrong with casting things in a more heroic mold. Disagreements like those between Mr. Duberman and Mr. Bell are fine. We need both kinds of teaching done with that kind of feeling in the university, but there is little of either.

Few people are actively building the kinds of communities that we want to see, and we are crucially lacking leadership. It does seem to me that for the universities to back out of a lot of activities is exactly the wrong way to proceed, precisely because the one great advantage we have is that students want to be whole people and to meet whole people; they want to be involved in lots of different things that challenge their minds and their feelings. That means there ought to be theater, music, and such things around. In regard to the problem of the university acting as an umbrella, it seems to me that the university can act more effectively if it shelters some things that the surrounding population is more happy with as well as some things that they may not like very much. We may be able in a pluralistic society to do both these things, whereas doing one and not the other would either be unsatisfactory ethically or simply impossible politically.

I think we should talk later specifically about such questions as: What do you do in terms of student life? What do you do in terms of making it possible for professors to teach, for building some kind of community on a commuter campus, on a campus where students are not involved in any formal organizations?

GRAY

I am impressed by the constant harking back to the past as a time when things were different. This is a nostalgia, a utopianism, that sees the past as having somehow incorporated values which we have now lost. In many ways the academic community did not possess those values in the past either. For example, the notion of a faculty that was once individually concerned with students as they are not today is wrong. The notion of a past in which there was a community of faculty and students that no longer exists is wrong. In the past, the criticism of a university that was being made by students was far less radical than now, but the students expected a good deal less of the university. Students became alumni; they then thought of the good old college days, of the professors who stimulated and inspired them, and translated these notions into the community that they had experienced. They then thought of that as the totality of what their university experience had been.

Although teaching roles have declined, although teaching is a lesser part of many faculty members' activities, the teaching that is done is without doubt better—better in terms of the standard and rigor of the courses. This is the consequence partly of the greater attention given to fewer courses, but also of real developments that have occurred in various academic fields. On the other hand, it is also true that teaching has become somewhat of an underground activity in the universities. Those who see many students on an individual basis are a part of what is a really new academic underground—the underground of teaching and of contact with students.

The quality of teaching has not declined, nor is the world of students more alienated than it had been in the past. The student feels that alienation, and his expectations as to what he is to derive from the institution have changed. That seems to me as important as the actual change which these institutions have brought about. These also seem contradictory desires. The desire to have a total community in which every kind of experience is open is directly counter to the notion of having an integrated community in which the members of that community are as one. It is like wanting to have a life of the village in the midst of a city, yet both students and faculty seem to be demanding just that. They want the whole city within the university and then complain when it turns out to lack the values of the smaller community.

There is an enormous gap between the rhetoric about education and the reality of the institutions which assert that rhetoric in their catalogues, in programs which they announce, and so forth. Universities are in a sense competing for students as well as for faculty. The gap between the rhetoric about what is to be made available and what the student actually finds is growing all the time. It does not just grow because professors are away in Washington; it grows because there is a real contradiction between the increased sophistication, the felt complexity of academic work and intellectual discipline, and the easy clichés in which all this is quoted. Not only does the institution promise the students something which it will not give them and which, if it is a good institution, it ought not give them, but the student is caught between the emphasis on education and the importance of professional training. One of the real lacks in our institutions now is not simply their structural difficulties; it is a matter of never saying what they are really about.

I am always amazed that students believe that so many of the faculty are on their side. The faculty is on their side; the administration is their enemy. To a great extent, these conditions are the product of faculty and of tensions that exist in the professional life of the faculty. Some faculty members are very good at seeing students in their homes, but that is not the solution to anything. Contact between faculty and students has to be made on those things with which faculty are most concerned within their own professional work, within the academic discipline. That contact has to do with the nature of the work that they are jointly involved in.

There is so little rational discussion of the complex issues that are involved, issues that cannot be simply absolved and where the conclusion of the discussion ought to be a further awareness of intellectual and academic integrity.

RIESMAN

Why in order for people to say that what is going on now is bad do they have to say it was once better? I have never been able successfully to combat this prevailing nostalgia. It exists not only about the university, but about the country as well. I have found no talk about the childishness and ham-acting of earlier academia. Students lapse back into the assumption that there were once great days.

GRAY

This clearly seems to be a repetitive pattern.

CONWAY

I think I can explain why those who are discontented with the modern mass university today do hark back to a golden past. Although it is true that what is taught in the modern university is conveyed with more intellectual rigor and discipline than, let us say, it was in the average college lecture room in the late-nineteenth or early-twentieth century, students today have many expectations concerning their university experience besides intellectual ones. Historically American institutions have been expected to serve a democratic end—namely, to multiply the options for academic and vocational advancement for all those who study or teach there. This function is, I think, still discharged well. What is new about the expectations which students bring to university is that they expect to find there the kind of environment for personal development which in the past was to be found at an elite college which followed the Oxford and Cambridge tradition. This expectation is not being met by institutions of higher learning today anywhere in North America. Yet I stress that this expectation is at the root of much dissatisfaction among students. I teach in Canada where most of the reasons for disenchantment with contemporary society to be found in the United States are not present, and yet we face the same kind of angry criticism of the university. Why this revolution of rising expectation should occur in the 1950's and '60's is difficult to explain, although one can sketch in the broad outlines of causation. Ours is a society in which affluence does permit for the great majority of students a concern with personal adjustment and self-development which was once possible only for an aristocracy. At the same time our culture is one in which there is a rising level of awareness of psychological thought. Philip Rieff has pointed out that just as nineteenth-century man was economic man, his twentieth-century counterpart is psychological man. One consequence of the incorporation of much popular psychology into our culture has been a change in the expectation of students about just what will be a good learning experience, and I think a corresponding change in the pedagogical goals of teachers. Many of us, I think, do see ourselves as teachers working to create a situation in which students will become self-aware and develop insight into their potential powers—not just, say, as students of history, but as whole human beings. We may not do very well at this, but to the extent that we do succeed in the context of a modern mass university, we are creating a situation in which students catch a glimpse of the possibility of self-development, while at the same time they realize that the existing occupational and economic structure of society is bound to frustrate that development outside college. Indeed to pass on to graduate school often means that the search for a fully elaborated consciousness must be abandoned for a personality structured to function well in some professional role. I think it is this profoundly disturbing awareness which is different from the past and makes people look back to a time when either the educational process was not consciously invested with such psychic significance, or when at least in some elite institutions the goal of personal development was a realistic one. The search for community is essentially a search for sharing of more than rational experience, and because of this it is not addressing the whole problem to discuss only the intellectual rigor of what is taught today.

PAYNE

In general, I think the hardline approach to the past is right. It seems to me, however, that the hopes of the students are the best thing we have, and we ought to be capitalizing on them. That we do not do enough. Of course, it is impossible to have the village in the city; the village has been gone for a long time. Nevertheless, it is not impossible to have a series of communities where graduate students, undergraduates, and occasionally professors talk together about the things that matter. That kind of community exists in places in Berkeley, so that it may be impossible to hire some Berkeley professors away, even though their academic situation is awful and the political situation worse. They stay at Berkeley for those communities. Those kinds of possibilities are clearly open. Students are not looking for the past, but for something that we have not had very much of yet. It is not just honesty nor just intellectualism. There are other parts of life, and those are not being met by universities now.

STOLZ

There is no better way to build a community than to have a generally approved riot. Everybody loves everybody else, and you get to know them; that's splendid. I am not sure that this a solution.

One of the mythologies of American higher education is that it has provided equal access to everyone as a way up through the community—a new route, a leveling technique where character and quality rather than wealth are the determinants of a person's capacity to get ahead. Going back to something Mr. Levi said earlier, secondary education is the place where we should be spending more money if our mythology of higher education does include this element of upward mobility. That mythology is sufficiently important and sufficiently a part of the tradition of American higher education to render silly the question: Should we have mass higher education or should we concentrate on graduate institutions? Politically that is not even a sensible question. We are clearly dedicated to the idea that anybody who wants a higher education and is capable of receiving it should get it. The problem is how do we do it, and how do we do it most efficiently. That is where I bump into what seem to me to be the critical problems—the economic ones. Where are we going to find the resources? The public sector is being squeezed hard, and there does not seem to be any reason to suppose that the situation is going to get any better in the future. Where are we going to get the resources to devote to mass higher education? Can we afford to indulge in some of the luxuries that we presently have? I do not know the answers to these questions, but I do feel confident that *we* are not going to make those decisions. They are going to be made outside the university. Unless we can build a persuasive case to the authorities of the public sector, we are not going to get the resources.

BELL

This raises a question which puzzles me. About four years ago, there was a fiscal drag because the growth of the economy was such that taxes were accumulating at a faster rate than the government could spend them. Now suddenly we find there are not enough resources. Which level is one talking about?

Governance of the Universities I

STOLZ

Jesse Unruh's Joint Committee on Higher Education in California obviously focuses on the resources of the State of California rather than the tax resources in the government as a whole, but it is very clear that the lush period in California's higher education was attributable to a certain accumulation during the war and the extensive utilization of credit in the period following the war. We have now run out of credit, and the demands of other portions of the state government are increasing, as are the demands within the educational establishment—including secondary and primary education. There is simply a squeeze on.

KAYSEN

This is partly a question of what perspective you wish to adopt at any moment. My own view is that there is a lot of money for education in the sense that the political process can easily and probably will provide a sufficient flow of money to do quite a lot. I think the acute questions are those Mr. Levi originally raised and Mr. Stolz has elaborated on. What particular packages for what kinds of things will look attractive enough to spend tax money on? In my own political perception, it is a mistake to compare this with the need for prisons or mental hospitals. The thing you compare it with is roads. We spend money like mad for roads because everybody uses them. More and more, higher education is coming in under this line. It will be used by a high proportion of the politically effective population of the United States. The people who do not use higher education are the objects of the political process and not actors in it.

RIESMAN

What about Negroes in the city?

KAYSEN

It is a mistake in proportion to emphasize Negroes in the city if we are trying to have a broad perspective. The parts of the political apparatus that deal with these things are completely different even though the programs and the agencies that hand out the money may be the same. In the one case, you are talking about a great political crisis and how we react to it. In the other, you are talking about how the government provides for something that everybody in society wants.

In 1945, there was not one analyst shrewd enough to predict the extent to which people in this country would want automobiles—not just one, but two or three. If this had been foreseen, we might have been able to come up with a better resource allocation. We built lots of roads because this was the easiest thing to appropriate tax money for; everybody liked it for a variety of reasons. I would guess that the whole education business is going to be that way. Whether it is fair or unfair (and Mr. Levi's observation that it is unfair is probably one I would share), the distribution of funds between higher education and primary and secondary education is going to be in favor of the former. We already have a set of institutions and tax processes that determine the resource size for the primary and secondary sectors. We do not yet have these for higher education on the scale that it looks to be growing. We will have to develop them, and it is always the new boy that gets treated the best. That is the way our political structure seems to work.

The question is not whether there are enough real resources in this country to buy this kind of higher education system, but how much the consumer will be willing to pay for it. And, of course, the consumer in this case is a most complex and curious congeries of state legislators, federal legislators, the executive branch, and various pressure groups. Trying to predict the outcome of this process is fairly difficult. One can, however, make some obvious predictions. It is going to be hard to say to a state legislature that a professor of Iranian manuscripts ought to be paid more than the governor. If the state is paying a man more than the governor to study old Iranian manuscripts and not to teach anybody anything, that fact has got to be buried in such a way that they do not notice it. It is not hard to persuade a legislature that a man ought to be paid more than a governor to teach us how to go to the moon, and because it is not hard to do that, you can pay people to read old Iranian manuscripts at these rates.

PARSONS

One of the most important conditions of the growth of the American academic system has been that the interests in it reach into wide sectors of the population through different functions, through different classes of institutions. Thus we have escaped the sociopolitical isolation that an elite system is likely to be exposed to.

KAYSEN

Competition alone was tremendously important; it is not an accident that the best university system in Europe was in Germany where you had the competition among separate sovereignties during the whole period of that system's formation in the first half of the nineteenth century. Competition among the sovereignties generated the competition among universities for people.

LEVI

A minor caveat—if you divide education between undergraduate and graduate, I am not sure that we do not have a good deal of an elitist student body. Most of our graduate students as opposed to our undergraduate students are very much like their fathers were when they were in college.

PARSONS

The undergraduate part of the system has been so strong and so viable to a great extent because there is by far the broadest base of interest on the undergraduate side.

KAYSEN

I disagree with Mr. Levi's caveat. He is in a law school, which is elitist. But in the natural sciences and to a great extent the social sciences, the affluence of government has had an enormous effect. Graduate students in these fields represent a much wider social spread than you would see in a law school because of National Science Foundation money. In the natural sciences it is relatively objective exercise to determine who is bright; the same is true in physics and economics. In a sense, by the time people are at the end of the first graduate year, you already know who is going to be a professor at M.I.T., Yale, Berkeley, Harvard, Chicago. On this level of the graduate system, I would say that there has been a great increase in the degree of egalitarianism; in the professional schools, practically none.

RUDENSTINE

We may be in this difficulty at the moment in part because the universities have managed during the last fifteen years to level compelling criticism at much of American society and indeed Western culture. This is certainly true politically. When a student arrives on the campus, he learns that we are in some sense an imperialist country, that our farm policy is out of phase, that our democracy is not precisely what he thought it was, and so forth. His educational experience is apt to be a fierce questioning of many of the values, political and otherwise, that he has accepted up to that point. The university criticizes our society and, therefore, gives students the impression that somehow the university and academics are better than the society.

This breeds extravagant hopes on the part of the students. They are inevitably disappointed when they discover that their professors may not act any better than the rest of the world. An issue like Vietnam makes that all too clear. Students discover—whether accurately or not—a tension between what the university says collectively in its books and lectures and what it seems to do at a moment of national or international crisis.

PARSONS

A large number of the people who demythologize the society hold out what I think many would regard as unrealistic hopes for the readiness with which society can be fundamentally changed. In other words, if they said what was bad about it, but also made clear how difficult it is to change it, their comments might have a different effect.

KAYSEN

If what Mr. Rudenstine is saying is the correct diagnosis, it is the result of some very bad teaching. Intelligent, responsible social scientists do spend a lot of time pointing out the way in which institutions and society do not function properly, how they fall short of their ideals, and so on. After all, that is what makes social science interesting and what attracts a great many people to it. But the same people, if they have any sense at all, point out that on any reasonable comparative standard this is one of the more flexible, humane, and open forms of social organization.

RUDENSTINE

I cannot help feeling that the emotional thrust of much of the work that has been done in the last ten or fifteen years has been deeply critical of America and its culture.

PAYNE

Students are always going to be fired up about issues of social justice, and it is not the teaching of the universities that has lighted those fires. In fact, it has in some ways been quite the opposite. A great deal of social science is devoted to justifying this political order, and, by the way, I think that is a decent task for social science. Students become activists partly because of the lack of a political understanding. The first time you face an issue of social injustice, you are going to be shocked, as people were shocked by what happened in the Montgomery bus boycott or during 1963 and 1964 in Mississippi. Those first political experiences are their first real perceptions of the extent to which the social order does not measure up to what they were taught by the Boy Scouts or in the first grade. That is where the ideas can be traced back to, although they are reinforced all the way through the system. When those ideals come up against harsh realities, the confrontation produces that kind of student feeling; it does not have much to do with what goes on in the universities.

GLASER

I attribute part of the difficulty to Mr. Kaysen's scientists who have inculcated us with the notion that truth is simple. Students get to college and, thanks to our educational system, learn that few things are simple. As a result, they find great comfort in the issues of social justice in which right or wrong can be clearly delineated. The issues of Vietnam and civil rights are simple ones for the students, as they are for most of us; there is a right and a wrong. In addition, numerous other matters, including the rebellious feelings that many current students have toward their parents and toward their elders, contribute significantly to current student unrest.

RIESMAN

I have been auditing courses at Harvard off and on for ten years. The audiences at Harvard are slightly different from those even at Yale, Stanford, and Princeton because Harvard students come from the sixth grade already slightly more cynical. The neighborhood is more tipped in that direction than it is in places that until recently have been slightly more square and are therefore more susceptible to activism and disillusion. In these classes, I constantly hear the refrain: Most of my colleagues are idiots; most of the books in this field are not worth reading, and yet you have to read them; most of the things that are believed in this field are not so. This tone obtains regardless of the professor's political position on the larger social issues.

In the general education class I direct, a colleague asked students who their heroes were, and they came up with, among others, Humphrey Bogart and Adam Clayton Powell. Although this is a coed class, there were no females. The feeling I have is symbolized by a talk I had with a Radcliffe student. She said: "The first thing I learned as a freshman in history here was that you got ahead by attacking the book and the professor. This defeminized me, and the problem of being a girl and being an intellectual was hopeless for me." This half of the university population is in a setting in the elite and potentially elite places where what they learn has a destructive undercutting element. It reflects the ambivalence of academic men—perhaps especially younger ones, but not all the younger ones—about what they do as well as about the society. One of the reasons why young women do not connect so much to the intellectual life is that they are both more responsive and more responsible than men, otherwise none of us would be here now. There is a real incongruity between the world in which they respond to people as well as to books and ideas and the one in which men respond by the narcissism of small intellectual differences.

KAYSEN

I have only been away from Harvard for two years after spending twenty years there. I do not know how Mr. Riesman goes to other people's classes; I cannot imagine anything more against the Harvard ethos than that. I must say, however, that in the department of economics, with a few trivial exceptions, you never heard a professor criticize his colleagues. If two colleagues had sharply different views, as they often did, you heard a lot of fairly serious discussion.

RIESMAN

Economists are, I think, more like natural scientists.

KAYSEN

Economists *are* more like natural scientists. I cannot imagine such comments in the natural sciences. The most I can imagine is conversations which say: "Oh, yes, X thinks thermodynamics is fascinating; I can't understand why he thinks it more interesting than field theory." It does not get any stronger than that. A few people are a little more passionate, but they are generally looked down on as being rather childish to talk that way. You do hear this kind of comment in a faculty of literature where there is an awful element of personal style. This is not just a sidelight that is personally interesting to us because we are professors, but something fundamental to what the academic enterprise is supposed to be doing.

GRAY

One would be far less likely to hear this among historians.

PAYNE

From the point of view of one who has been a student over the last eight years or so, the kind of thing that Mr. Riesman mentions is not the universal experience, but it is the most common. It is the most common subject of graduate student discussion. I have heard it very solidly in history departments and throughout the social sciences and the humanities.

RIESMAN

Wouldn't you rather agree with me now, Mr. Payne? The thrust of my statement was that students are taught to repudiate. The implication is that somebody who teaches you to repudiate somehow ought not to be repudiated.

PAYNE

I think this is true at the academic level; I was only questioning it as a political explanation. I do not think it explains much about student politics.

MORISON

This nostalgia versus hopes or hopes versus nostalgia is a critical issue. At our college, maybe 15 per cent of the people come with hopes which are frustrated, and they take it out in activism and so forth. So purely from that standpoint, it does seem worth thinking about. I think they are hoping for specific sorts of techniques and information, and universities are far better able to give them that sort of thing now than they were when I went to them. You can be better prepared for a profession; you can be better prepared and discover truth on your own. Those hopes seem to be satisfied. The hopes for some meaning in life are not being satisfied, but in my opinion they never were and the university was never designed to answer such needs. The question is whether you redesign the university so that it can take the place of the other mechanisms which man has had in the past to make life a meaningful experience.

In the twenties at Harvard, there was no personal criticism of colleagues on the faculty, but there certainly was a great deal of skepticism. The whole purpose of the enterprise was to shake the beliefs students came with; we did not feel so upset about this. Somehow we were quite pleased to discover that there were people around who would question those rather rigid value systems in which we had been brought up. The Boy Scouts were very real for us in those days, and many of us went to church. Our parents were quite clear about what they thought life ought to be like and what we ought to be like, much clearer than parents are today. Consequently, we found this skepticism refreshing. I have a feeling that the students who are most upset now are the ones who do not come to a college with values—either from their parents, their church, or the Boy Scouts—and are looking to the college to give them a set of values and beliefs and meaningfulness which universities simply have not been designed to do for a very long time.

RIESMAN

We are the gods who failed.

MORISON

Yes, and I am not sure that I want to do anything but fail.

PAYNE

There are going to be more and more of those students coming.

MORISON

It is really a question of whether a university is a good mechanism for erecting a positive value system to which everybody can tie themselves.

PAYNE

But there is a second question: Is a university a good place for getting a hold on that kind of framework?

MORISON

Since the discussion about the role of the teacher as a role model, I have been jotting down certain thoughts. It seems to me that there is a hierarchy in this. I would say that the biologists, chemists, physicists, and literary critics, in about that order, are pretty good in supplying role models. There are creative artists around universities, but they do not show what the role is in that line. We do a poor job of providing a role model for businessmen and political and government figures. We tend to deprecate those to the extent that the student feels that that cannot be the way to engage in society. The difficulty is that most of the things that go on in society are carried on by either businessmen or politicians. We do not seem to have been able to give any dignity to that very large sector of society which is engaging in society's problems directly. The student rejects these roles and tends to feel that the university ought to engage itself directly in social problems. I am not sure that that is quite the thing. The university is perhaps failing to show that the people who do engage directly in society have something to be said for them; that it is not so bad as it looks on the outside.

PARSONS

When I was an undergraduate, the business community was very much debunked. I was partly concentrating in economics at that time, and Veblen was the hero of my principal teacher. He did not give a damn for Alfred Marshall or anyone like that. Veblen was his hero, and Veblen was about as destructive a social critic of American business as you could find.

MORISON

Few of my classmates thought business was a good thing to go into.

WAX

Although very few of your classmates thought very much of business, it seems they went into it.

MORISON

Right.

WAX

My impression today is that the bright people who are coming out of good universities are not only opposed to business, but will not go into it. That is why Motorola has this big advertising campaign in the Harvard *Crimson*; obviously there are people who go into business, but the larger percentage of them do not.

KAYSEN

Does somebody know the facts? My own impression was that the proportion of college graduates going into business is about the same now as it was in the twenties.

PARSONS

No, it is decreasing at the elite colleges.

LEVI

I am terribly skeptical about information about students. There was a time, however, when it was not so upsetting to have students upset. Today, the notion that somebody is frustrated or upset seems to demand some kind of response. I am not sure that it warrants that kind of response. Being upset is probably a good thing in the educational system. A lot of the criticism of the kind of education given obviously comes from those schools where the education is very much the best. There seems to be a lack of urgency, a feeling of boredom, a grasping for even greater luxury, and a very peculiar value system. A foundation made a study of the present college situation, and part of the grief they discovered arose because the rooms for college students were not very nice. When one thinks of the pressing problems of our society, it is immoral to center on this kind of thing for a luxury class.

I do think that the university should stand for something. But what? I do not think it should be deceptive. It should stand for intellectual truth as it sees it. I completely disagree with Mr. Duberman. There is a sense in which finding one's self intellectually is a part of the whole thing; you do not have a disembodied intellect sitting there anyway. It seems to me that the important service, the real task, of the university in this country is to provide intellectual truth. I do not believe that it is fulfilling this role very well. The social scientists are very much to blame for this. They have inculcated the notion of deception, rather than rationality, which is extremely dangerous. The humanities, philosophy departments, and divinity schools have added the notion that intellectual honesty is impossible for a lot of reasons. If the universities do not stand for intellectual integrity, they are not properly communicating with their students. The judgment I would make on a university, then, is whether its size and the activities that it takes on are such that they defeat this basic conception of the university's purpose. I realize that what I have said is very foolish, because there is not one kind of university. Part of the problem is the terminology; we ought to have different words for different institutions. We ought to separate them out. There is a whole variety of institutions. One of our problems is that we keep talking about things over one ball of wax, and we do not separate out the talk about universities within the university from the talk in *Life* magazine and over the television.

It is terribly important that the institutions find their own integrity. If it is not the rational intellect as the main focus for some institutions, let it be something else then. Nevertheless, this problem of purpose has to be met head on. Part of the problem of undergraduate education is the problem of graduate education. You cannot justify a great deal of graduate education. We have much too much of it, particularly in the humanities. We have erected an enormous structure in this country, and we find ourselves discussing whether there will be sufficient financing to keep it going. I doubt whether there will be, and, furthermore, I think it is improper to ask for that financing. Many other needs are much more pressing in this country, and it is wrong to have this enormous structure of education which cannot refuse the challenge to take on everything and loses its identity as it does. The kind of university that I am talking about would have to keep its size down to a basis where it can have an intellectual mode and mood and candor which go throughout the place; there can be other kinds of institutions. Not only would this be good for the institution, but it is imperative for the society.

The whole urban redevelopment field is a dramatic illustration of much of what I have been talking about. Social scientists do not know how to write about it, but they write about it, and generally speaking with enormous inaccuracy. They remake history gladly and happily, because they know that that is what history is for. They do not have a professional sense of responsibility. Quite apart from that, they are not the people to do it. There is the notion that we turn to the universities because that is where you can tuck the problem. What has been done is to blunt the charge of society on these points.

It is curious that we maintain this elaborate graduate center when there are enormous jobs that could be done on the outside. We lock students into graduate activity for years on the theory that everybody has to get the doctor's degree. Not only that, we have told them that they have to do it in order to make money, in order to be trained. There are other ways of training, and adult education is very good too. Moreover, we have locked *ourselves* into a multilithic-monolithic structure of what we think a university is. This is the problem.

BELL

The important prior question is who should be educated and how far? What standards do you have to determine this? How does one create such standards? One of the things which has bothered me is that about 50 per cent of the students in most state universities drop out by the end of the second year. These universities have the habit of taking everybody who maintained a C average in high school—and in many cases not even a C average. Madison, Wisconsin, has been blown up completely out of proportion because it now has forty thousand students and will probably have forty-eight thousand in a few years. This has literally wrecked the university; half of the students are deliberately dropped by the end of the second year, which is an enormous waste of resources.

DUBERMAN

If you reduce student discontent to how large their dormitory rooms are, of course you trivialize that discontent. I, however, contend that this discontent is so important that it weighs equally with what is happening to our cities or to our Negro population. This generation of students is not content at age eighteen to be told that somebody else has the right to make the essential decisions in their lives. They feel more mature than that, and from my observations they are more mature than that. In other words, they are protesting the whole super-structure of *in loco parentis*, their lack of control and involvement in basic decisions that have touched their lives.

In asking that we consider or re-evaluate the role of teaching, I was not necessarily implying that the re-evaluation would involve an increased amount of time or work or expenditure. Such an implication would come very badly from me, as I teach at Princeton and live in New York City. We need not so much to expand the number of contacts between student and faculty, but to look at the quality of the contacts that already exist. We need not to add ten hours to the teaching load of each professor, but to re-evaluate what is going on during that initial ten-hour period. Our investigation might in fact lead to a reduction in expenditure. By a variety of definitions, student-initiated and student-run seminars are often a better form of education. This would mean an actual reduction both in faculty teaching levels and in expenditures of faculty time. We might want to do away with certain kinds of classes altogether, whether or not they were run by students or faculty. We might think that certain kinds of material are best dealt with outside the university structure.

GLASER

Should students be intimately involved in appointments and promotions?

PAYNE

Yes.

GLASER

What expertise do they bring to the selection of a professor of biochemistry or of Latin?

PAYNE

At institutions where teaching makes a difference, where there are standards that say teaching is important in terms of the appointment of a faculty member, students want to know whether those standards are being followed. At Yale about three years ago, we were not sure that the official standards of the university were being followed. A student who sits on that committee during the course of a year can say, in general, whether a person's teaching ability has been taken into account. There are lots of sources for information about teaching ability. For example, at Yale some of the top seniors and the top freshmen are asked to describe their educational experience. The question that is raised is whether those opinions get into the discussion. If they are not getting into the discussion, something is wrong. In some departments, in some tenure committees, they do not because people try to maintain an objective stance and shy away from the evaluation of teaching. Students do not need to have a veto power nor a controlling power of any kind. They do need to see the process as it goes on. There is no reason that representative students cannot know about confidential matters. The more confidential information that students have, the more responsible they are likely to be; and conversely.

GLASER

I gather Mr. Payne believes that students will not accept the assurance of the faculty that these matters are considered critically and thoughtfully. If they did, they presumably would not be concerned. Assuming that one student was on an *ad hoc* committee of five people to consider the promotion to tenure rank or the appointment of a professor, how long do you think the students would be satisfied if they found that their criteria for this appointment and those of the faculty did not coincide? Inevitably in many areas there is going to be great stress put on productive scholarship. How long would the students be satisfied with that system and not want to take over the whole thing?

RIESMAN

At Harvard, and the same may be true at other leading institutions, the students internalize faculty values very early. They make judgments of faculty harsher than the president, certainly harsher than the trustees. I would worry about their conservatism.

PAYNE

The suggestion that there would be an *ad hoc* committee on questions of tenure or appointments is probably not a good one. I would want to institutionalize that kind of committee. If a department discusses appointments, students ought to attend each of those meetings. Over the course of a year, you would find that the disagreements would probably be fewer and fewer. If there were serious disagreements, however, they ought to be aired. If students sat in on more departmental discussions, in a great many institutions their fears would to a certain extent be laid to rest. If it turns out that a problem is one of personal clashes within a department and not a question of teaching and research, it is not bad for responsible students to know that. These things have to be discussed; eventually they are discussed and known in any case.

DUBERMAN

The problem often is not whether standards are being followed in regard to teaching, but whether such standards exist. The value of having student representation would be that perhaps for the first time in the university we might get a general discussion as to what good teaching really means. Students can tell us better than anyone what strategies and techniques do or do not succeed.

PARSONS

I would question the assumption that students are the best judges of teaching. There is the old legal principle of not making a man judge his own case. The student has a definite self-interest in the teaching, but it need not be wholly objective.

Governance of the Universities I

PAYNE

I would not say that students are the best judges, but they do provide the information on which the best judgments can be made. Students understand perfectly well that if a professor is a good teacher for 20 per cent of his students, he may be very valuable to keep around even if he is not a good teacher for the rest. Channels of communication might get much better with student representation on such committees.

DUBERMAN

Many of the suggestions that some of us have made have been regarded in the nature of "instead of" suggestions; they ought to be seen as "in addition to" suggestions. I am not saying that students by themselves are proficient to judge teaching, but neither are professors by themselves. I should like to see student opinion added to that of departmental representatives.

MEYERSON

The etiquette of American colleges and universities is such that we have no source of knowledge about teaching other than students; it is the rare institution where we attend one another's classes or in any other way have a sense of how our colleagues teach, what they teach, and their effectiveness. In that kind of unfortunate etiquette, if we do not lean upon students, we are being foolish. By saying "lean upon them," I do not necessarily mean that they have to have the fifth seat on an *ad hoc* committee.

It is critical that we provide ways for student opinion about teaching to be heard. We must make sure, however, that we weigh that opinion. Max Planck never had a large number of enthusiastic students, but he should not have been dismissed because he did not succeed well in that particular popularity contest. Such cautions have got to be built in, but we also have to establish some channels whereby student opinion can be heard. Few universities have those channels today.

KAYSEN

The distinction between an organized way in which student opinion can be taken into account and student participation in a decision is quite important. The arguments for an organized channel are interesting and worth considering. Systematic attention to teaching ability has merit.

I should also like to say something which will not be received sympathetically, but is still important. It is extremely difficult to make an appointment, which is why it is difficult for a good department to remain good. Students can be trusted. They would be no worse as sources of gossip than faculty. Nevertheless, it takes a long time for the sense of shared values to grow up among a group of people which permits them to talk with some candor and to expose their foolishness. I have heard my distinguished colleagues say things which I would never quote back to them because they are so stupid. They say them in the heat of discussion, because they do not want a particular man appointed.

If a student were on the committee, the discussion would not proceed. The faculty are not going to trust him. He is around for three years. Faculty contemplate spending on the average twenty-five years with their fellows. They adjust their thoughts, feelings, and interpersonal sentiments to make that possible. It is hard; often it does not work. Many good departments tear themselves to bits.

One of my colleagues at the Institute says that there is an exponential law of faculty: First-rate people appoint only first-rate people; one second-rate person leads to a fourth-rate appointment which then leads to an eighth-rate appointment. He exaggerates a little, but this process is extremely complicated and difficult.

I once tried to explain to one of my colleagues in Washington why Harvard professors were so reluctant to resign when they came into government service. I pointed out that anybody who has been through the appointment process once knows how difficult it is. This is not a process which can be exposed to an essentially unserious influence. The great work of a university is intellectual. No one who has not made that commitment—and very few graduate students have made it—is entitled to participate at this level. That issue is quite different from the question of how you get intelligent and organized student opinion about who is a good teacher. Universities do this very badly.

PAY.

Having a student on a committee ought not to make it impossible for the discussions to proceed. I admit the difficulties of keeping a first-rate faculty, but students today would be willing to put that amount of time in. They are serious about what they know. Harvard and Yale make a lot of bad appointments, occasionally on the basis of teaching and particularly at the lower levels. Sometimes they appoint a professor who is supposed to be a good teacher when his teaching is not that good. As Mr. Duberman said, we are asking to add something to the process. The difference between having organized channels to discover student opinion and having a student sit in on the deliberations is that in most places students simply are not going to trust the organized channels unless the channels reach to the top. They may at Harvard; they do at Yale. I have been in many difficult discussions, and I do not see why one other person, even if he does not understand what is going on in some ways, makes it impossible to say certain kinds of things.

KAYSEN

I predict that if students come on, the faculty will go off.

DUBERMAN

What is being said is indicative of the basic failure of student-faculty relationships today. Faculty feel they can be vulnerable, make mistakes, and expose parts of their personality only in front of their peers. Students are human and, in most cases, mature—often as mature as their counterparts on the faculty. If faculty have trouble saying certain things in front of students, they should be forced to confront their trouble. There is so little communication between faculty and students today because faculty members will not permit students to see them other than as masters, as perfect creatures.

CONWAY

In our discussion, we have been talking about stable institutions in which the informal power structure has been legitimized by custom and habit, where informal consultation could go on, and where it is perfectly clear with whom you should consult. That is not true of most state institutions that are growing rapidly. I think benevolent despotism is wonderful if it is really benevolent and genuinely efficient. You have it at Harvard, and it has functioned marvelously. But I should like to examine the analogy in a little more historical detail. Benevolent despotism worked well in the small German states, but when the attempt was made to apply the methods of enlightened absolutism to the Austro-Hungarian empire, those methods were disastrous. Despotic methods cannot be applied to the government of a large lumbering institution, particularly one which is undergoing rapid change.

We have not talked much about the question of instability and growth, but I think it is very important to do so. At the University of Toronto I have become involved in all sorts of decision-making processes simply because the administrative structure there was designed for a small college, but the institution has now grown large. When a university is expanding rapidly there are many power vacuums; there are many instances in which administrators are simply not there to make appropriate decisions, and so many administrative tasks fall back upon faculty committees. A faculty member at such a growing institution is in a political situation whether he wants to be or not, because power is thrust upon him. Another problem of institutions experiencing rapid growth is that they do not possess an informal power structure known and accepted by the whole university community. It takes time and familiarity for such a structure to establish itself. We should view the question of participation in terms of developing some new kind of legitimacy for the decision-making process, and remember that it is critical to develop this legitimacy in new and rapidly growing institutions. We might also ask ourselves how fast institutions can grow and remain viable, because I think there is a point in institutional expansion where change is so fast that it becomes impossible to define a stable power structure or establish any basis of legitimacy for authority.

GRAY

Obviously universities are oligarchies; the question is whether they are restrictive or open oligarchies. Even if there can be no question of absolute democracy in a university, there can be constitutional monarchy. One is essentially talking about how this oligarchy becomes more or less democratic.

Although we constantly use the analogy of a political system, the representation characteristic of political systems is not possible in a university. A professor's professional commitment is such that he is never going to let himself be represented by somebody else. A faculty representative asserts his own views, sometimes taking the trouble to find out whether these are also the views of the people whom he represents. On the whole, however, he argues a case as he sees it in the same way that he would argue a case in the abstract or if he were simply individually asked. Thus, the notions of democracy and representation in a university have to be qualified by these obvious differences between the university and a political system. These differences come out of the structure not only of universities, but of intellectual life as well.

Furthermore, the ways in which the political structure at universities has grown up are such that it looks like a rational matter. Everybody knows exactly what the pattern is and is able to suggest the way in which one could tinker with it and make it still more rational. Most questions then become procedural ones. But underlying this passionate rationality is a structure which is traditional, which has grown up in a Rube Goldberg way.

How does one preserve the individual emphasis which is part of the intellectual life of a university together with the belief in the need for some kind of consensus, for majority rule, for a stricter representation than is possible within a university?

While faculty at some institutions are clamoring for more participation, at others where faculty participation has always been great, they are withdrawing. It is difficult to get people to sit on committees at such places. They are withdrawing in those institutions, but wanting at the same time to be consulted on matters of individual interest to them. There cannot be, obviously, any solution to the division of functions within a university. Universities are reluctant to abolish anything. They build theologies all their own.

RUDENSTINE

The primary question is whether the oligarchic leadership can be good enough to establish a regime of trust and confidence that will make students believe that they do not need to take part in selecting tenure faculty.

LEVI

You cannot go through a situation such as the one at Berkeley without making mistakes. I am not impressed by the predictions. It may very well be that universities will be ruined. There is no reason, however, for us to be on the majority side, no reason for us to participate in ruining them. Many universities may not be ruined, but they may become even more mediocre than they now are.

It is odd that people say that universities have the capacity to change society, but do not have the capacity to resist the evil influences within themselves. I am not impressed with these arguments, nor am I impressed with the claims of the seriousness of the situation. A university president is used to resisting pressures. I do not understand why he should not resist them.

There is a great danger due to governmental involvement in the universities. It might be assumed that because government is the all-embracing thing of life, every part of life should be in the same model as the government, that everybody should participate in all the decisions.

I am most serious about the university and its purposes. If you change those purposes sufficiently, you change the institution. It might go on, but it would be much less interesting to many of us. The consequence would be the creation of other kinds of institutions.

If the politicization goes to these extremes, I assume that institutes will develop and that the graduate and undergraduate areas will become separate. I am not sure that would be bad, although it certainly has not been the direction that many of us thought we should go in. Nevertheless, it is a likely outcome if Mr. Payne's predictions prove to be true.

The civil rights tactics have been adopted by students, and the response of younger faculty who have their own problems is important. There is a group of students who want to destroy the university; they have written documents pointing this out. The most important thing that a faculty does is to make determinations on appointments and promotions. It is difficult to force someone to speak critically about these issues. If these become public confrontations, the process would entirely change. The only consequence is that more mediocre people would be appointed. The institutions that allow extreme student participation will become mediocre; maybe all institutions will, and others will have to be created.

I do not like the language which says that it is quite all right to speak about the intellectual purposes of a university, that all we want to do is add other things. That is very seductive. If you add on things, these additions change the university's purpose. There is an enormous amount of participation in a university. In many institutions the choice of a dean involves formal faculty participation. Nobody is opposed to turning over to students a variety of things.

The younger generation may, as Mr. Payne predicts, radically change the universities; it may also have to explain to its children how it happened to ruin these institutions. If such is the case, I would expect other institutions to grow up, other institutions which are interested in intellectual strength.

MORISON

I should like to add a few footnotes to Mr. Levi's comments. Since I have had my present job, we have lost five tenured people, and four of them went to nonteaching institutes. This could be coincidence, but I am not at all sure that it is. They left because they did not want to participate in all of the things one presumably has to participate in these days. They did not want to do so much teaching as we thought they ought to do, even though this was not very great.

This flight from the university is disturbing, and I was thinking of it earlier when Mr. Payne commented that knowledge about teaching is not taken into account enough in making appointments. It is not purely a question of information. I can tell you quite accurately the teaching ability of at least seventy of the seventy-five people in my group. I interview about two hundred students a year to see whether or not they want to major in biology. I ask them informally how they are enjoying their courses, and they tell me. We started a new course last year with four of what we thought were our best teachers, and it turns out that they get standing ovations at the end of every lecture. These people are people I want to keep.

The question is how you use this information to get something done. Many people at Cornell know that certain elementary subjects are badly taught. These courses infuriate the students. I do not have to have students on a committee to know this. Nevertheless, it is extremely difficult to persuade a department to include some people in it who can teach elementary courses. This becomes an administrative or management problem.

It takes a lot of time to use this information. It is easy to find out what is lousy in an institution, but it is difficult to do something about it. I do not quite see how students can help in this. Students have legitimate gripes. I do not think our teaching is so good as it should be. We rely too much on graduate students and do not get the best team in the first-year courses. It requires infinite tact to get first-rate people into the first-year courses. If you put too much heat on them, they go to an institute. Most administrators are scared to death that they are going to lose the people they have. Anybody on my staff who is any good at all is getting between two and five letters a year. The amount of heat you can put on a faculty member to do what you know has to be done is limited. If administration gets any more complicated, it will be increasingly difficult to get anyone who will administer.

PAYNE

I am sure that Mr. Levi is going to be able to resist almost any kind of pressure. That is not the question. My suggestion was that we could make a few creative responses to the student anxiety beyond simply resisting. There are adaptations that can be made. I was not suggesting that students ought to be involved in all the difficult tasks of administration. I made a specific proposal that on one of the questions that bothers the students most—the question of tenure appointments—they might have some guarantee that the information that ought to be coming up through the channels is coming up and is being heard. I do not know any way to get that guarantee other than a feeling of confidence in the administration and faculty, so I opt for a structural innovation. Structural innovation is possible; it has been suggested not only by myself at Yale, but by the dean of the graduate school at Harvard. I find faculty at Yale not nearly so frightened about it as Mr. Kaysen.

KAYSEN

It will not do any good.

PAYNE

It may very conceivably do some good in terms of the crisis of trust and confidence that I was talking about. I grant it probably will not result in the appointment of many more people who are that much better.

BELL

The words "student participation" always bother me, because you tend to think in terms of numbers. Is this supposed to be a symbolic gesture or an actual participatory gesture? You have thousands and thousands of students, and only a few are going to participate even if you succeed in all your structural innovation.

PAYNE

If those students are talking to one another and are involved with a group like the Student Advisory Board at Yale, which is concerned primarily with educational policy, these checking mechanisms are probably adequate. If something seems to be out of line, that issue can be discussed not necessarily in terms of specific people, but in terms of the general direction in which the university is moving.

KAYSEN

Are these undergraduate or graduate students?

PAYNE

I am thinking primarily about undergraduates, but my argument would apply equally to graduate students on certain other issues.

Governance of the Universities I

DUBERMAN

I am interested in the university continuing to serve the function which theoretically it has always served—a devotion to intellectual pursuits. The university is not now succeeding in these pursuits; it is certainly not communicating enthusiasm for those pursuits to its students. It is not succeeding mainly because of the separation that exists between faculty and students. Rather than student participation corrupting the current aims of the university, it might for the first time help to fulfill those aims. All kinds of inhibitions against understanding could be broken down by an increase in trust and confidence between students and faculty.

PAYNE

The question is not whether a single decision is made on which there is disagreement, but whether there is a pattern of such decisions. In many instances, students do not have any recourse until an extreme case develops and they can blow it up publicly.

RIESMAN

I served for a year along with Dean Monroe on the Harvard policy committee, an educational group selected by the residential Houses, therefore violating student power groups who wanted campus-wide elections which would allow political parties to form. Mr. Payne should be more familiar than most of us here with the agony of such meetings. There was seldom a quorum, people came late and left early. They were frustrating meetings, and yet it was an enormously worthwhile group. It took its work very seriously, negotiated with departments about curricula, and got through a pass-fail system. In working with this group, I was constantly caught between the faculty right and the student left. The student left regarded this group as a co-opted body. Even at Harvard, which is relatively small in terms of the Houses, there was not much communication between the student body and this group, who by and large took their task very seriously both in the subcommittees working with individual departments and in the committee as a whole. They got little support from fellow students; they were, in fact, much attacked as stooges.

One of the things that this group could potentially do if this is an area of concern would be to try to free the vision of arrangements from the parliamentary model. This mechanism at Harvard would have worked better if people had not thought it evil *per se.*

MEYERSON

I should like to second Mr. Riesman's comments. They are extremely important.

BRADEMAS

What troubles me about this discussion is that a choice seems to have been imposed on us between student participation or no student participation. The same point could be made, for instance, with respect to the role of the trustees. It may well be that students have no business expecting their voice to be taken seriously with respect to the choice of a new president or a new dean, but perhaps in respect to other decisions something can be said for listening to what they have to say.

Nobody has as yet addressed himself to the question of who speaks for the students. I turn your minds to the community-action programs and the poverty war wherein we had written into law a requirement that there be participation on the part of the poor. The problem of trying to decide who speaks for the poor is a much more difficult problem than we had anticipated. Do the students have elections on campus? Are their leaders self-appointed? Are they the heads of the local university student associations? What if there is a fight among the students as to who speaks for the students? What kind of mechanism for deciding do you develop? Who decides the issues on which student opinions are to be taken seriously? It ought to be possible to have informal mechanisms as well as highly formal ones for enlisting student opinion.

The Federal Government is getting into this act. If I am going to be voting, as Mr. Pifer suggests, for 50 per cent of the monies that a college president is going to be using, what about my participation? What do I have to say about what goes on at a university since my vote is going to determine, in part, 50 per cent of the money that the institution may get? To turn this question around, since the universities will be directly affected by congressional legislation, ought they not to have something to say about that legislation and about the mechanisms Congress develops to ensure that Congress hears accurately and effectively from higher education in the United States? What, in fact, does it mean to say "hearing from higher education"? If you are talking about a national policy for higher education, how do you decide which student leaders or which university presidents are going to be heard? Think of the proliferation of individuals and institutions in American society that are directly affected by what higher education does. How will their voices be heard if we are writing legislation that will have such a tremendous financial impact on American higher education? This discussion has not yet addressed itself to what may be a far more serious problem than student participation in terms of policy-making for American higher education—that is, what does the government have to say?

GLASER

I want to illustrate what Mr. Brademas said. Because the curriculum at Stanford is changing so radically, we decided to have student representation and discussions. We asked that the students give us two people to participate in a group of eight or ten. We left it entirely to the students to determine how they would do this. I was soon visited by a small group of students, representing perhaps 5 to 6 per cent of the total student body; they announced that they did not approve of the representatives who had been selected and that they would consider any participation by them in the curriculum discussions as invalid in respect to their own personal interests. They are not prepared to accept anything in which they themselves are not involved. If their classmates do not decide to make them representatives, they say the procedure is invalid.

RIESMAN

I am much in sympathy with what Mr. Brademas said, having watched the land-grant and the private colleges battle these issues, and having felt the appeal of institutional grants to almost all administrators. These grants are not in the interests of students, but in those of the going concern. This is one of many reasons why I favor the Zacharias plan. It seems to me to be a counter-weight to the institutional grant. The institutional grant involves the university in on-going, but not always happy, momentum.

BRADEMAS

I was in Iowa the other day talking to a group of presidents of small private colleges. Their conversation started off with an attack on the large land-grant institutions. This attack was not, in my view, a particularly useful enterprise. One of the ways in which the dilemma has been put is this: Assuming that there will be a substantial increase in federal support, should that support be provided in the form of general institutional grants or in the form of a variety of categorical aids, as is presently the case? There is a similar battle going on at the elementary and secondary school levels. I happen to be militantly opposed to the general grant approach at the elementary and secondary school level for a variety of reasons which need not be aired here. I am not honestly sure, however, whether or not the same kinds of arguments obtain at the level of higher education. As one prejudiced in favor of categorical grants, what does concern me is that if we were to provide general monies to every institution according to formularized percentages, the grants would not amount to much. You might see an inflationary consequence financially, rather than an increase in quality in respect to the purposes for which categorical aid is supplied. At the very least, higher education in this country has not thought through the problem of financing higher education.

Within the last several weeks, I have begun to hear university presidents and others saying that their institutions are going to go under financially in about ten years unless something revolutionary occurs. These comments are, I suppose, triggered in part by the proposed cutback in facilities money under the higher education bill and also by the cutbacks in money for research. You can normally predict a land-grant college president's response in respect to student-aid proposals. A private university president, for example, is likely to opt for some tuition tax credit approach, where the other fellow is not happy about that since tuition is not his major headache.

I strongly agree with Mr. Pifer's suggestion that we develop some kind of mechanism for shaping a coherent and national policy for higher education. I am not suggesting a highly centralized mechanism whereby Congress dictates what universities do and do not do. Nevertheless, with so many different institutions of higher learning of so many different kinds in the country, all of which will have an increasing stake in the decisions taken on federal money, we must develop some way of deciding who gets what and for what purpose. If we do not do that, we ought to at least consider the issues involved. If we cannot establish some kind of procedure, we ought at least to talk about the critical issues. At the present time, it is a nonsystem in which everybody goes his own way in some general *laissez-faire* approach.

KAYSEN

Mr. Brademas, at what level would you judge that the situation will get so rough that the Congress will be willing to do some unpleasant things? We have a large number of small colleges. They do not educate a great proportion of the student body, but they do educate a not trivial proportion of it. They are inefficient, and, with few exceptions, they are terrible. Haverford is suboptimal in size, but it is quite good because it happens to be rich. How much money would you have to be spending before you could write into a bill a statement that says no dollar of student aid could be used in an institution that does not have X thousand students, assuming that you had competent testimony that X thousand was the reasonable floor for an efficient institution?

BRADEMAS

There is a certain unstated assumption in your question: namely, that you close up the small colleges rather than make the small ones good enough to survive. That is an open question.

KAYSEN

Let's assume that you cannot make the small colleges good enough unless you have fewer of them, which would, in effect, make them larger. What do you think is the point at which Congress will get tough in the sense of demanding a show of resource? The present programs, even though they are fairly large, are not big enough to exercise that kind of pressure. Every special interest group can get in and get something in the program because it is reasonably sized.

BRADEMAS

Your question illustrates the point that I was trying to make earlier. When you have two or three thousand colleges and universities of every size, shape, and description devoted to different kinds of educational purposes, what does the phrase "to get results" mean? Some qualification is necessary before one can make an intelligent judgment on how much money would have to be spent.

There are other policy questions that at least ought to be thrown into the cauldron. In the South there are a number of struggling Negro colleges, and there are also barely viable institutions in Appalachia. On a cost-effectiveness basis, I have little doubt that you could make a case for closing those places forthwith. But what do you do then? Perhaps in some cases it is better to have second-rate universities than none at all.

KAYSEN

Suppose you were to conclude that a few of these places should be closed so that the others, especially the ones that are in or near cities, could grow. You might let the country community college, the country Negro college, and maybe the country Appalachian college just die in order to provide not fewer total places, but a different institutional distribution. When the defense budget got so big, we became less sensitive to certain kinds of pressure groups.

BRADEMAS

I do not know the answers in monetary terms, but I should think that we are going to be getting to that point within four or five years.

KAYSEN

There will be a point when it is no longer possible for Congress to accommodate whoever comes in with the combination of a plausible program and a constituency.

PARSONS

Let us hope it is not an across-the-board formula, but a highly selective program.

BRADEMAS

Who is going to make the selection?

PARSONS

I think it has to be a collaborative governmental and academic body.

BELL

I do not see how Congress can escape making categorical judgments on higher education. The Zacharias plan, although useful in providing flexibility, does not take into account the total picture. It does not face the primary problem of the categorical judgment. The educational plants in this country are largely in the hands of the states because of the distribution of constitutional responsibility. The states have become increasingly inadequate as an instrument for carrying on education.

If you begin to look at the whole federal scheme, at what might be called regional institutions and regional needs with cluster arrangements, the policy of giving monies to states makes no sense. They simply follow existing lines and probably reflect the political weights of the different states. If there is going to be some distributive justice, one cannot wholly concentrate on the state unit. One would have to take into account certain kinds of regional distributions.

Perhaps a different kind of cluster arrangement might be viable in which you begin to distribute certain kinds of strengths or insist upon the distribution of certain kinds of strengths. A regional cluster might allow for distributive strength so that people in the area could share resources in a particular range. In that case, there would be a great responsibility on universities themselves to begin to think in terms of sharing their resources rather than in terms of the higgledy-piggledy pattern which is now in existence. Presently, universities simply grab faculty and follow the free market at its worst in this regard.

One would also want to get a clearer sense of resources. We do not know what the complete demands are, but we do have a fairly clear sense of what the resources are in terms of specialists in Latin, in Russian, and in various other areas.

BRADEMAS

I agree. I want to endorse thoroughly what Mr. Bell has said. One of the parts in the higher education bill which we are probably going to finish up next week is the so-called "networks for knowledge" title which is aimed at encouraging precisely that kind of inter-institutional cooperation. Although we have not got any money yet for the International Education Act, it has at least had the value of encouraging a number of institutions to make an inventory of their libraries, students, and faculties.

MEYERSON

I am worried by Mr. Bell's last comment and Congressman Brademas' agreement with him. Most efforts in which the Federal Government manipulates coordination have been dismal. We have seen such difficulty in federal stimulation of metropolitan planning, and I can see some of the same failings occurring in higher education. We have seen problems in the national systems of education in other parts of the world and in the state systems of our country. In California, as an example, the decision was made that African studies were to be concentrated in Los Angeles even though most of the students interested in African studies might prefer to be at Berkeley.

This tendency for federal coordination may come up with extremely conventional approaches. A safer approach is the one that

Congress has traditionally followed. Congressionally the United States has favored a series of competing claims and different viewpoints. We may even have had contradictory programs running side by side. I do not demand coherence.

I would hope we would have not a carefully worked out national Cartesian scheme for higher education, but rather a series of competing and even contradictory programs. We would thus avoid the problem of imposing a pattern that may not be a good one. It is also important to have the source of financing going through the students so that the student as a consumer of education can make his choices. The existing programs ought to be continued and amplified. There ought to be other programs too. I can, for example, imagine branches of the Library of Congress throughout the country serving the academic community; instead of asking for two copies of a copyrighted publication, the Library of Congress might ask for two dozen. This would be simple to do, and the Library of Congress could put the other copies in large regional depositories.

There are many approaches of this kind. It would be far wiser to diversify the sources of support than to try to rationalize them. In diversifying these sources of support, it is important for universities to make sure that they have private as well as public support. We ought to try to develop a mixed economy for colleges and universities.

BRADEMAS

You exaggerate my point greatly if you assume that I am in profound disagreement with a great deal of what you have said. Nevertheless, one must still come up with a generally intelligent national approach to legislation supporting higher education. This does not necessarily entail a centralized Cartesian rationale being imposed from on high.

BELL

I agree with the second part of Mr. Meyerson's comments, but not with the first part. With the second part, he ends up with a mixed economy; with the first part, he wants *laissez faire*. This is a crucial distinction. Clearly none of us want patterns; we have all had enough experience with that. We may, however, want to have some sense of the guidelines, some sense of planning, without patterns.

Society, as Mr. Meyerson remarked earlier, has made a decision that everybody should have the opportunity for some higher education. Society did not make that decision because there was a conscious plan wherein somebody said X number of persons a year will be allotted to higher education. There was a *laissez-faire* situation and a rush in terms of aggrandizing in many states. California had some degree of planning; many states did not. Along the way, many institutions became wrecked or are in the process of being wrecked because there was no notion of how many students a year they could absorb, no conception of the costs of such a pace.

The Governance of
the Universities II

With these few caveats, the following agenda was proposed for the discussions on November 14-16 at the House of the Academy.

American colleges and universities are today, as in the past, immensely diverse in their organization and purposes. To pretend that the problems of the public junior college, concerned with the "career programs" of nonresident students, are fundamentally like those of the great private or public university, with its continuing commitment to undergraduate and graduate instruction and to the advancement of research, is to ignore substantial differences in favor of superficial similarities.

Still, to see no similarities in their situations is to forget how much the United States is becoming a "national" society, subject to common pressures. Greater numbers of young men and women expect to be accommodated in colleges or universities. There is a prevalent belief that a high-school diploma does not provide sufficient training for the most desirable kinds of employ, and that a thirteenth, fourteenth, fifteenth, or sixteenth year of schooling ought to be made available to high school graduates. No one seriously contests the "right" of young people to these additional years of education, and the sentiment in favor of making places available for those who wish to fill them is not likely to diminish in the near future. There is an analogous opinion that advanced training in the professions is another obligation that cannot be turned away from without seriously harming both the individual and the society. The "right to higher education" is seen very differently today than it was even a few decades ago.

The increase in the number of students, independent of any other variable, guarantees that higher education will remain a heavy charge on the nation. Although institutions may seek additional funds partly from private sources, both public and private institutions will certainly require vast new appropriations, and these will come largely from public sources. As both private and public institutions turn increasingly to state and federal agencies for support, they must be prepared for a type of public control and audit that has rarely existed heretofore. Arguments about appropriations and currcula, not to speak of those that touch student and faculty conduct, will almost certainly become the subject of public political debate. The old "independence" of higher education will be seriously affected by these new public pressures. The conflict between a public demand and a university's own sense of what it ought to be doing may become very intense. Each may be interested in innovation, though each may wish to define that possibility differently. Within the university community itself there may be real differences in the definition of appropriate service.

In recent years, universities have substantially increased their research activities and dramatically revised traditional nineteenth-century definitions of university "service" to the community. The prospect is for society to look increasingly to the university for various kinds of assistance, thus perpetuating the current stress within the university over the priorities assigned to teaching, research, and "service." The contemporary debate about the university's role as a critic of society will certainly continue. The guaranteeing of certain rights in this area and the fulfilling of certain obligations become matters of urgency if criticism is an essential function of the university, and if, as some suggest, the intellectual authenticity of the university is tied to its capacity to protest.

The ambiguity that has long existed about whether the university ought to be viewed as a sanctuary, isolated and protected from the general society, or as a microcosm of the larger society, responsive to its needs and pressures, is debated more passionately today than at any time in recent history. If the university is seen as a monastery, one set of relations may be expected to prevail; if it is viewed as a small city, other kinds of relations will be valued. Perhaps the university ought to resist all efforts to make it appear analogous to other kinds of institutions; it is neither a church, a city, nor a business corporation, though it may bear superficial resemblances to each. The need may be for a more fully formulated expression of what, for lack of a better term, might be called "university law."

This "university law" would seek to establish the rights and responsibilities of the various constituent elements of the university. It would aim, among other things, to describe the kinds of authority that ought to subsist in the university and to create principles of accountability that would protect all who have a stake in the university.

The American Academy's effort, if it is not to traverse too many fields, might reasonably address itself to the following five topics:

1. The rights and responsibilities of trustees;

2. The rights and responsibilities of university administrators;

3. The rights and responsibilities of faculty members;

4. The rights and responsibilities of students;

5. The rights and responsibilities of the general public, as expressed particularly through the state and federal legislatures, but also through alumni organizations.

There is a hazard in a too formal distinction between the rights and responsibilities of these various constituencies. The discussion ought certainly to be informed by a precise consideration of the kinds of issues that presently preoccupy university communities. Thus, for example, one of the crucial problems of any institution is whether or not it ought to take on additional functions. These are various and may include anything from service to the federal government in the field of defense to involvement with local communities in urban planning or secondary-school curricular revision. Who is to decide whether or not the university becomes involved in such matters? Until recently decisions of this kind were generally made by administrators in consultation with trustees. Should the faculty have a voice in such decisions? Of what kind? Ought students to be consulted? On what principle?

Clearly, the purpose of the Academy study is to ask: How are important decisions made today in higher education? What are the sources of dissatisfaction? What might be done to resolve these discontents? The conference ought to direct its attention to the larger policy questions presently agitating universities—whether they touch on the draft law, the use of university facilities by other than university members, investment policies, or the like. It would be a mistake to neglect the more strictly university issues that have to do with course offerings, housing arrangements, and the regulation of student activities.

In thinking about each of the five constituencies, one ought always to have in mind specific examples. The conference must think not only of the issues that presently "make the headlines," but also of those that are likely to confront universities in the future. Some of these have not yet become urgent, but they may soon figure prominently in public debates on higher education.

Agenda

I. *Rights and Responsibilities of Trustees*

A. What ought their appointive power to be? In most colleges and universities, trustees have for all practical purposes abdicated their traditional right to veto the appointments of faculty (when voted by departments and approved by administrations). They retain large (sometimes, exclusive) authority in the choice of presidents. Should this authority be shared? With whom? On what principles?

B. Their fiscal authority, legally and actually, remains substantial everywhere and is sometimes absolute. Should it be limited in ways not common at this time?

C. What other powers inhere in boards of trustees? How should these other obligations be defined? Is there a hazard in boards of trustees assuming that every matter affecting the welfare of the university (issues ranging from student discipline to the acceptance of architectural plans for new buildings) properly falls within their competence?

D. What kinds of administrative regulation should not be considered the responsibility of trustees?

E. Is the principle that trustees ought to act as a buffer between the university and other bodies in society a reasonable representation of their proper role?

F. To whom ought trustees to be accountable? How can such accountability be provided for? What moral and ethical considerations ought to guide trustees?

G. Is it possible that the concept of the "trustee" is outmoded and should be set aside? What would be gained or lost from such a change?

II. *Rights and Responsibilities of University Administrators*

A. If the administration, particularly the president, is accountable to the trustees for the governance of the university, what kind of audit ought to be instituted so that the exercise of authority is not judged simply by how the administration comports itself in time of crisis?

B. What relations ought to subsist between administrators and faculty members? What controls may the former reasonably have over the latter? What restrictions may faculty members reasonably impose on deans and presidents?

C. What relations ought to subsist between administrators and students? What sharing of authority ought to be provided for? How can this be institutionalized?

D. Can the responsibilities of the administration be effectively discharged when so many in the administration are faculty members who hold office for relatively brief periods? Ought there to be a larger career opportunity for those in universities who permanently choose to be administrators? How can they avoid being thought "second-rate citizens" in the better universities and prevented from appearing to be "the controlling force" in the lesser universities?

E. What rights ought administrators to claim vis-à-vis state legislatures, alumni groups, and other public and private bodies that have reason to be concerned about the university?

F. Are there any areas where administration ought to claim exclusive authority? Where ought administratration to be ready to share authority? How can systems of accountability be constructed that will not hamper effective action, nor stifle initiative from constituencies outside administration, both within the university and without?

G. What values ought to enter into the administrator's decisions?

III. *Rights and Responsibilities of Faculty Members*

A. What are the obligations of a member of the faculty to his colleagues? What are the faculty member's obligations to students? Is the idea of the corporate faculty disintegrating, giving way to a too exclusive concern with professional and other allegiances? What hazards are posed by such developments?

B. What rights ought faculty members to enjoy with respect to the appointment of their own colleagues and the creation of suitable curricula? Are departments, as presently constituted, a barrier to educational innovation? How serious is this? Does departmental organization provide positive benefits? What are they?

C. What examining prerogatives ought faculty members to enjoy? Ought these powers to be shared with others? On what principle?

D. Ought there to be any boundaries governing intellectual dissent? Are the foundations of the university jeopardized by attempts to impose such limits?

E. What concerns ought faculty members to have with the "needs" of society? Should these be reflected in a willingness to alter curricula to serve those needs? Also, does the present commitment to accept extra-university appointments imperil in any significant way the freedom of universities?

F. What are the research obligations of faculty members? What liberties must be guaranteed if those obligations are to be discharged?

G. To whom ought individual faculty members to be accountable? Are there special obligations that properly inhere in the role of professor and need to be acknowledged? Is it possible that the socialization of young men and women into this role is happening too quickly and too haphazardly? What obligations do older (tenured) faculty members have to those who are younger (presumably, serving term appointments)? How can faculty members express and communicate faithfully the personal as well as the academic values of their profession?

IV. Rights and Responsibilities of Students

A. Is there some area where students ought to enjoy exclusive authority? What ought that to be? Where should students share authority with others? On what principle? Are the present conventions governing student rights in respect to curricula outmoded? What control, if any, ought students to have in this regard?

B. Does the status of student impose obligations? What ought they to be? Who is to judge whether or not they have been transgressed?

C. What political freedoms ought to be guaranteed to

students? Ought they to be different from those granted others?

D. Have young people a "right" to a university education? Does this mean that the traditional methods of student selection and admission are untenable? How ought they to be modified? By whom?

E. Should the fact that students are still overwhelmingly adolescents or young adults, passing only a brief time in the university, be given any weight in determining their rights and obligations? Ought differences to be established between undergraduate and graduate student rights? Why should all not enjoy the rights that might be granted a postdoctoral fellow? What ought those to be?

F. What claims may students legitimately make of their professors? What may they reasonably demand of university administrations? What, in turn, may professors, presidents, and deans ask of students? Does the status of student carry with it certain rights? What are they?

V. *Rights and Responsibilities of the General Public, Federal, State, and Local Governments*

A. As major benefactors (and beneficiaries) of the university, what are the legitimate rights and obligations of government—federal, state, or local—and of alumni organizations? What kinds of control must they explicitly forswear if the university is not to be deflected from its purposes? What kinds of accountability may they reasonably demand?

B. What budgetary decisions ought to be made in the political arena? Is it possible for this decision-making process to be institutionalized in such a way that it is not directly tied to politics? Is it desirable that there be a separation between the two?

C. What restraints must the general public accept in respect to its right to interfere with the criticism made by faculty members and students of institutions (and, often, of themselves)?

D. Does the existence of the university in a democratic society, pledged to giving opportunity to all, impose special kinds of obligations, both for the university and for state and federal legislatures? What are these?

E. What information ought the university to provide about itself so that those outside may be in a position to gauge accurately its needs and its performance?

F. Which of society's needs must the university be ready to acknowledge and serve? How can this service be guaranteed? What respect must the university be prepared to give society's opinions? What is society's responsibility in defining the demands it makes upon the university?

Participants
Landrum R. Bolling
Peter J. Caws
Sarah E. Diamant
C. M. Dick, Jr.
Seymour Eskow
Edgar Z. Friedenberg
Stephen R. Graubard
Andrew M. Greeley
Jeff Greenfield
Eugene E. Grollmes
Gerald Holton
Willard Hurst
Dexter M. Keezer
Clark Kerr
S. E. Luria
Jean Mayer
Walter P. Metzger
Martin Meyerson
Robert S. Morison
Henry Norr
Talcott Parsons
Bruce L. Payne
Roger Revelle
Philip C. Ritterbush
Neil R. Rudenstine
Edward Joseph Shoben, Jr.
John R. Silber
Charles E. Silberman
R. L. Sproull
Kenneth S. Tollett
George R. Waggoner

MORISON

I have a predilection for thinking about what a university is for and whom it serves. Indeed, I am not sure how far we can go inside the university without thinking about those on the outside. But before my prejudices show too clearly, let's turn to the agenda and begin to discuss trustees.

HOLTON

I wonder whether we should not perhaps first define the framework. The subtitle of this conference is "Study in Academic Ethics." We do not want to lose sight of that aspect. What, then, is the role of the trustee not with respect to day-to-day things, but with respect to the main theme of the conference?

DICK

Perhaps my experiences as a trustee will help start off the discussion. Three years ago, I became a trustee of a liberal arts college with an enrollment of about five hundred. A friend asked me to come on the board, and I agreed to serve because I am interested in education. My friend had been chairman of the board; when he resigned, I took over the chairmanship. A lot of important things were going on at that college—problems with the students, faculty, and president—and I got an extraordinary education in a short period of time.

Speaking from my own experience, I would say that the primary responsibility of the board of trustees is to keep the institution going. Our particular board came up against this point time and time again as the future of the college became the principal issue. The second responsibility of the board of trustees is to provide guidance and policy to the administration. The president, to whom the board has delegated operational responsibility, should expect this contribution from the board. The third responsibility is to try to insure that the financial resources are available to enable the college to carry out its program.

MORISON

To whom do you feel yourself responsible?

DICK

We certainly are not responsible to any particular body that comes to our meetings and says: "You are responsible to us." We say on various occasions that we have responsibility to the students, to the faculty, to the other board members, to the president, to education in general. The thing that keeps the board together is that all its members want to see the college keep on going and improve. I cannot think of a better way to define the responsibility.

SILBERMAN

I wonder if there is anyone here who is a trustee of a state university or municipal college. It seems to me that the agenda itself betrays the initial bias of the steering committee—a tendency to think of the university as the elite, private university. The representation here is very heavily from the private universities, although the great majority of students are now attending public institutions of one sort or another. In a sense, the dichotomy between the trustees and the public is broken in public institutions where at least formally, if not substantively, trustees are assumed to be the representatives of the public. If you ask the trustee of a state school to whom he feels responsible, he would answer that he is responsible to the public or the legislature. It might be useful in terms of bridging these few items on the agenda and perhaps of dealing with the question of purpose if one could get some sense of how trustees of public institutions view themselves and how this view differs from that held by trustees of private institutions.

KERR

I am not a trustee at the moment, but for a period of time I was a regent of the University of California in an *ex officio* capacity. I could make a distinction between what trustees should be and what they really are. What they really are varies enormously, depending upon the nature of the institution, its history, its traditions, and the particular composition of the board at a moment in time. Trustees also change depending upon the issues with which they are faced. I do not know whether we should discuss what trustees should be or what they are.

Charles Silberman raises an important question. Are trustees representing the public to the institution or are they representing the institution to the public? When the chips are down, that question is absolutely central. In the case of the University of California, when the trustees and regents were not under pressure, they thought that they were representing the university. When they were under pressure, a lot of them decided they were representing the public.

Governance of the Universities II

PARSONS

There is a group that, curiously, does not figure in the agenda. You might call it part of the not-so-general public—namely, the alumni. Alumni figure rather prominently in the affairs of many institutions of higher education. The extent to which trustees or administration or faculty are responsible to the alumni raises complicated questions. The alumni ought not to be left totally out of the picture.

FRIEDENBERG

I wonder if it might be helpful to try to be slightly more functional in drawing inferences about what trustees do and to whom or what they are responsible under what circumstances. It has come to be customary to select trustees from within a relatively narrow and familiar range of other social roles, which may provide a reasonably valid clue to the questions we are asking. The trustee's task is almost always a part-time and usually unremunerated one in this society. Trustees are primarily drawn from either successful business ventures or in some cases the bar. Very rarely are they drawn from scholarship or the arts, which are more usually the main concern of the university's curriculum. I take it that the selection process has to do with the trustees' having a kind of alarm or fusing function—that is, they are expected to remain inactive and do remain inactive until certain kinds of interests, generally upper-middle- or upper-class interests, come to be frightened. Surely it is proper for a trustee to be devoted to the continuity of his institution. The ease with which we accept that premise may have much to do with what we have been calling the crisis of the university. Indeed, one of the questions very much before this meeting is whether it is desirable that the university be kept going if it is to have such a preponderance of the legitimate social roles of young adults in the United States. If the only way you can keep the place going is to have a system where the final panic button rests in the hands of such people as normally become trustees, then I rather think I should like to see funds going elsewhere. Some of us yearn for an irate public that would throw us out into experimental colleges that could not grant degrees, could not contribute much to vocational education, and therefore would not be besieged by the lower-middle classes simply because these settings could not do them very much good. We might get the equivalent of the suburban shift in education where we have to build our own poor utilities and live less lavishly to begin with in order to have certain kinds of freedoms. If we look to the charter and ask what a trustee is supposed to do, then I think we will be way off. University people are not any different from any other segment of the population when they start talking about law and order.

HOLTON

We must distinguish between governance and representation. Absentee governance seems to me an ethical problem whether it is in slum housing, in a family where the mother is not at home, or in those universities where the governors are essentially absentee landlords. In the latter case, faculty might be regarded rather like tellers at a bank.

Indeed, much depends on the models that exist in the minds of a university's governors. In the modern governing board of a university, physically and professionally distant from the real "business" of the university, the model for running the university may well come from some more congenial activity such as the local bank, the law office, or whatever other field of enterprise the trustee is engaged in. In such cases, there is a premium on keeping the thing going more or less in the same way year in, year out. On the other hand, in my laboratory the whole object is to terminate an experiment successfully and in an orderly manner to go to something entirely new and different when a brighter line of research opens up. The Cambridge Electron Accelerator is not going to be kept going at all costs and through all crises forever. Some day a quite different facility is going to be put in its place when its present usefulness is over. Also, in the internal governance of a lab, the voice one has in the long- and short-range decisions is much larger than the voice students and faculty have in the governance of most universities. The danger I am pointing to is that the governance of the university by predominantly business-enterprise-oriented groups may impose upon the university institutional models which will inevitably get the institutions out of step with a changing situation.

Neither the lab by itself nor the business corporation by itself may be the best model for the governance of a university. The trick is to produce the right model intermediate between both to assure stability and change.

RITTERBUSH

It seems to me that the central problem of competently functioning trustees is to guarantee that the university stays governable. We should bear in mind that in the corporate sector there are well worked-out procedures for going through transitional phases—such as bankruptcy proceedings—which do not serve to extinguish the array of talent that has been assembled to do a job, but do establish it on a new footing. Institutions should not necessarily fear entering a spore-forming phase to knit themselves together in some new way. If universities and other institutions do face a time when major change is needed, it may be desirable to shut them down for a year or two, indulge in a period of patient reconstruction, and then open them up again. If there were somewhere in this society where a group of conscientious trustees, for there are many such, could turn for advice and help as to how to go through this almost clinical experience with their institution, many of them might be encouraged to try it.

BOLLING

One of the threads running through our discussion is a tacit assumption, which you find on many college campuses, that the trustees are a necessary evil, but unmistakably an evil, and the less they are heard of the better. The concept that a significant group within any institution is only a window dressing—a tolerated nuisance—means that you do not get the best possible results out of that group of people. We have not given nearly enough thought to the question of how you select trustees in the first place. What kinds of people ought to be on boards of trustees? We have not given sufficient attention to how to inform trustees and keep them involved. One of my presidential colleagues once said he thought he had an ideal board of trustees because they only met once a year for a half day, heard him make reports, patted him on the back, and went home. He did not want any more involvement than this. I think that is nonsense. I think also the question comes up as to what the real role of the trustees ought to be. This again is not clearly defined. We do not want them meddling and trying to overrule the administration and prying into the affairs of the faculty, and so on. Yet they must have a significant role in the over-all policy-making of the institution, or they are going to be frustrated. Otherwise, you will only get people who are collectors of titles or honors to do this kind of job.

GREENFIELD

Why don't you want them meddling in what the faculty is doing?

BOLLING

The trustees ought to be informed and involved and selected from a sufficiently wide range of competences in our society that they have useful opinions about the total operation of the college. It seems to me that one of the most critical difficulties about educational institutions today is that we have not gone beyond the concept of just keeping things going the way they are going. We are not taking a hard enough look year by year at what we ought to be doing. Who is going to do this? As a one-time faculty member who now has the prejudices of administrators, I would submit that the faculty are jolly well not going to do this. Some group that has the over-all view of the institution and of the society, one that can break loose from its departmental hangups, has got to take on the job and keep the pressure on the institutions to reform themselves.

There is a mythology in our society today that the teaching faculty is the great fountain of reform and change. This is just nonsense. You have got to have an informed board of trustees and a sensitive, informed, and determined administration, at times backed up and allied with the students, to change the educational institutions so that they will be what they ought to be in our time. Here the trustees have got to be an important participating element within this total effort to rebuild and renew.

The question of institutional renewal is certainly one of the most crucial issues that we face. When you talk about governance, you have got to talk about not just governance in terms of keeping things going, but governance in terms of making the institutions relevant to the needs and problems of the times. I think this is where the trustees can be very useful—if you pick them correctly, if you inform and educate them, if you give them significant things to do. Then I think they can be a very creative force; otherwise, it is going to be terribly frustrating for everybody.

MORISON

Are you and Mr. Greenfield ganging up to say that the real engine of reform and change in American education should be the boards of trustees as such?

BOLLING

No, but if they are properly selected and informed and involved, they can interpret to the public, to the donors, why reform is necessary, why it should be supported.

GREENFIELD

We have always thought of a university as something like a corporation in that it is separate, in the private sector. If universities are going to get more than half their funding from public institutions, if they have the kinds of public effects that we know they have, it may well be that there are going to be constitutional requirements—much like the one-man, one-vote decisions of the Supreme Court—demanding that universities, both public and private, have effective, meaningful representation from all sides of the community. I am talking about judicial decisions based on the Fourteenth Amendment. I am saying that if Columbia University, located in the heart of Harlem, does not have a single black man on its board of trustees, that comes close to an unconstitutional use of power against the community.

RUDENSTINE

Mr. Greenfield is right, I think, up to a point. Boards do have to be democratized for many reasons. At the same time, he is a bit sanguine. It is clear that when "democratization" happens, there will be many practical implications and repercussions. It will certainly mean that black people and other kinds of people will come on to boards, as indeed they ought to have a long time ago. But as a result there will be new cleavages, tensions, and important changes in universities. Many people will resist such development, and that will create even greater tension. I fear that the academic community as a whole is simply not looking at certain problems that are going to be with us in much more chaotic ways very quickly. I am dismayed, for example, that many universities do not yet have people in the administration who work on community problems. It seems to me that until universities face up to their public roles, they will continue to be in great trouble.

At the same time, by abstracting the role of trustees and either hoping for too much from them or blaming them too much, we are overlooking the problem of the academic profession itself. Mr. Greenfield suggested over coffee that the trustees should tell the faculty that they could not expect any further salary increases or special kinds of privileges. But the problem of high career expectations—financial and otherwise—is clearly a national cultural one and the faculty cannot be held entirely guilty for its particular expectations. We will not solve such problems by having the trustees tell faculty that there are not going to be provisions for housing and other fringe benefits or any more salary increases.

LURIA

It seems to me that the basic issue is who ultimately will set the purpose of the university. It is certainly not the trustees who set the purpose. The point that Mr. Rudenstine and Mr. Greenfield made is excellent—we are visualizing a dynamic process in our society in which the purposes of the university are being reviewed actively under the leadership of both students and special community groups. The only way that we can expect universities to respond to the situation is by creating, within the university, machinery that continuously re-examines the question: What is the purpose of the university in this specific situation, in this community, in this state, or in this type of technological demand? I should like to propose the following: The best function of the trustees, if they continue to exist, is to steer the machinery of the university by responding to the community. A board of trustees should be a plastic group maximizing the inputs and providing the machinery for the response. One of the problems has been that trustees in the past inevitably have acted as instruments to slow down processes of change. On the one hand, they have represented the need for efficient, financially sound operation and, on the other hand, they have been drawn from a part of society that traditionally has not been actively interested in change. If the trustees are to continue to exist, they must take the leadership in making the university extremely responsive to new pressures from within and from without.

GROLLMES

I would suggest that the role of the trustees is going to vary depending on the particular kind of institution. A trustee at a large state university is going to have significantly different concerns than the trustee of a small Catholic liberal arts college. It is important that the trustees understand not only what adjustments the school must make in order to make its contribution to contemporary society, but also the purpose of that particular school, especially in the sector of private education. Private education, it seems to me, is based upon a presupposition that private institutions offer a particular kind of education that you cannot get at the secular, state institutions. Thus, one role for the trustees is to insure that the institution be true to itself as well as to its public. In being true to itself, if Shakespeare is right, the institution will serve the public.

One of the recent phenomena in Catholic higher education is precisely in the area of trustees. Formerly, with very few exceptions, the top executives in Catholic colleges or universities were also the trustees. There really was no distinction so far as personnel were concerned. As they have brought in lay people and religious people from other institutions, they are finding that the trustees can serve as a check on the executive. When the executives were the trustees, there was no one to say that the institution was not doing what it said it was going to do. I think it is fundamental to the success of higher education that the trustees not only have an understanding of society and what is needed today in the whole realm of higher education, but that they also evaluate the precise function of their particular school.

NORR

It seems to me that our analysis of the functions of the trustees—the ways and means by which they should have a role or whether indeed they should exist at all—has to be seen in light of an issue that has been mentioned, but not really dealt with here: that is, the sociological and constitutional source of the authority of the trustees. If the trustees are to continue to represent primarily businessmen and corporate lawyers—and not the students and faculty or certain members of the communities around the universities—then most of us would feel that the thing to do with the trustees is to ignore them or minimize their role or, best of all, abolish them. Their abolition would involve looking toward the creation of new mechanisms by which the interests presently not adequately represented in the governance of the university could be represented. If, on the other hand, we admit that we can amend the concept of what a board of trustees is—sociologically and constitutionally—then we can indeed talk about ways in which the trustees can and should act. If we talk about the kinds of changes that would be necessary in order to make the trustees into a body that would represent the community, the students, the faculty, the general public in a broader sense, the issue of how the trustees can act as a check on the selfish interests of other elements in the university becomes more meaningful. But I do not think it is worthwhile for us to talk about the rights and responsibilities of the trustees without asking who they are, where they come from, what the source of their authority is, what motivates them to devote their time, energy, and resources to the university, and what interests they in fact serve in running the university.

PARSONS

I think Mr. Norr raises a central set of questions. I know the faculty is not our current topic, but Mr. Rudenstine brought up that constituency in the context of the professional role, as individuals and corporately, at several different levels. If it is correct (and there may be a good deal of discussion on this) that the primary functional focus of the system of higher education centers on knowledge—the advancement of knowledge, the transmission of knowledge, and the maintenance and use of the cultural traditions of knowledge—then from one point of view the key role of responsibility is in the academic profession. We are all well aware of the extent to which the professional privilege can be abused. Mr. Greenfield called attention to one type of abuse; but the pressure exerted by faculty for improved conditions is not a simple one. Faculties have had too much of a "stand pat" tendency to preserve the status quo. If, as Mr. Bolling suggests, trustees combine with administrators to put pressure on faculties, if severe pressures develop to force faculties to change their ways—especially in directions that deviate from the primary concentration on knowledge—then I would join Mr. Kerr in predicting the large-scale unionization of the academic profession. Faculty could become very obstreperous indeed from the point of view of their relations with trustees, administrators, and possibly students. I do not think, therefore, that you can leave the professional concerns, standards, and traditions of the faculty as a secondary aspect of the general problem.

WAGGONER

I am struck by the readiness with which we, in our closed system of higher education, accept the board of trustees and then start tinkering with the system in a pragmatic way, as if that tinkering were all that is necessary. Clearly, the public university stands in a somewhat different relationship to organized democratic society than does the private university. Mr. Dick described a concept of the role of the trustee in a private university: to make policy for the university and delegate the administration of that policy to the president. In most of the public universities those are exactly the legal powers of the board of regents or the board of trustees. On the whole, this system has probably not worked too badly in the United States, but I should like to take a few minutes to relate an anecdote that I think leads to a generalization. A couple of years ago I entertained a group of foreign rectors and deans for six weeks. I asked them to evaluate the State University of Kansas in the method of our U.S. accrediting associations. They observed things that normally we do not observe and focused on things that normally we do not. First, they wanted to read the statutes of the university. They observed the legal role of the regents and were shocked. How could a university community turn over policy-making powers to a group that was completely nonacademic? This seemed absurd to them. The American concept

Governance of the Universities II

of a board of trustees in public universities is a rare thing in the world. It is not the traditional method of administering a university. They went on to say that even though shocked, they were compelled to admit that it seems to work pretty well. How could this be true? Obviously it can only be true because there is total unity between the governing group of society and the aspirations of the university. When no conflicts arise, university autonomy is not a significant matter. I met a Canadian Communist on his way back from Cuba a month or two after that. I asked him about university autonomy in the University of Havana. He said that was a silly question: When you have an absolute unity in society and an absolute purpose, you do not have to bother about a university being different. Instead of a board of trustees for a public university, why not put it under the state office of education, an agency of the people and its processes, or under the ministry of education, as has been done in the Communist countries?

I balk at comparing boards of trustees to board of directors of corporations, administrators to corporation executives, and faculties to the production-line workers in a factory. The student, of course, is the fabricated product. This does not work very well. The evidence is that the student does not behave like an automobile coming off the production line. A traditional view of the university in the whole history of Western education would be that the university has a special kind of role in society and that it demands independence and autonomy. The last thing that one would want would be this constant nonacademic definition of the purposes and policies of the university. I would speculate that the board of trustees is probably an anachronism carried over from the private, small liberal arts college of the nineteenth century where one board of clergymen had to supervise another group of clergymen who were teaching and were not really much to be trusted in their dealings with students.

MORISON

Are you *for* or *against* the autonomy of the university?

WAGGONER

I am for it. What I am arguing for is a concept of the autonomy of the university as a community made up of students and scholars and teachers which does not need an external policy-making body and cannot accept one over the long run and fulfill its function as a university.

MORISON

Would you say that the Latin American and European universities have more or less autonomy?

WAGGONER

Many Latin American universities have much more. They have problems which are sometimes blamed upon their autonomy, though I myself would not accept that reasoning, and would attribute their problems to other factors.

LURIA

Is this not a critical issue? If a board of trustees has a representative role in the sense that the society has a real voice—whether it be the internal society of the university or the larger society—could the board effectively mediate so that the problems the Latin American universities meet would not be encountered? Cannot the trustees play a buffer role? Performing that function, the board would not prevent shock so much as mediate between the university and the society. It would determine what the university thinks it wants and test those proposals for feasibility within the university. An intermediate body of people can play such a role as long as it is not representing only one small sector of the community.

CAWS

As Mr. Holton predicted earlier, the discussion has kept pretty much to structural considerations with a few moral interjections now and then. But we are commissioned to talk about ethical questions. Meetings like this are called, I think, because of a sense of moral crisis in the community at large. It is important to realize that one of the unclarities in this situation is who, precisely, is the agent. Moral questions arise when people do things that affect other people or fail to do things that affect other people. If we are to talk about morality in the university situation, we must have some conception of who is acting. I do not find it helpful to use simple collective terms like "the faculty" or "the trustees."

Mr. Bolling said a little while ago that change is not going to come from the faculty. I certainly do not think it is going to come from the trustees as boards are presently constituted. Thus, the question is: Is there a recognizable collective agent? There is one obvious sense in which the corporate model is appropriate: The trustees originally constituted the legal entity that was empowered to receive funds, own real estate, and have responsibility to some constituency or other. They were collective agents. The issue confronting American higher education and perhaps higher education around the world is whether these entities as presently defined are fulfilling their moral obligations. Are there actions that the university performs or fails to perform vis-à-vis the society at large that it ought to be performing or not performing? The Columbia trustees are a case in point. If one asks a moral question about agent, action, and consequence at Columbia, it appears that there has been something lacking in that situation. Should some sort of evolutionary descendant of the present board of trustees continue to be the agent that has responsibility toward the internal and external community? Or are there other models of the collective agency of the university that might be just as good, if not better?

Mr. Waggoner also said something which struck me. There is a well-known tradition in certain places that faculty members and students are in fact members of a collective, members of a community, with legal standing. They are not simply clients or employees; they *are* in some sense the institution. I do not see why one should not explore seriously the conception of the university as a collectivity having certain members. I should myself wish to restrict membership in the university to students and faculty, who would then hire the administration. The administration would be a service, not quite on the level of the janitors but having a rather similar relationship to the body itself. Students and faculty together might very well elect a body that took on their behalf the collective stance of responsibility toward the society.

I thought that Mr. Greenfield's remark about the adversary role of the trustees was admirable. The faculty has never been tried in this connection. Within faculties, some of the best ideas are very often generated; yet the present governing arrangements do not make it easy for those members of the faculty to be heard. We have not yet even begun to explore alternative structural possibilities in the light of the moral challenge that confronts higher education at the moment. I must align myself with Mr. Friedenberg's demurrer to Dr. Morison's summary in this respect. I do not think we can assume that a reformed board of trustees will do this job. We may have to turn to a collective agent of an entirely different nature.

BOLLING

I suppose I was playing devil's advocate; I frankly shrink from neat categories and any attempt to approach these issues by separating out the faculty and the administration and the trustees.

You cannot brush off these problems in terms of a bureaucratic tidying-up or a slightly reformed board of trustees. We need a new concept of ultimate authority within the university community that will embrace more fully than it has in the past the teaching faculty and students. You have also to bring in some representatives of the general public to look at what we are doing. I do not think there is any special virtue in any category of people. We all partake of original sin in a sense. I would not trust the faculty and students any more than I trust the administrators or the board of trustees to have the exclusive authority. You have various interests that have to be looked at, and there is something to be said for bringing together people who will have influence upon one another in weighing what is being done. They must work together to produce devices, mechanisms, and policies for renewal. I would not for a moment suggest that the trustees should be put into adversary roles, although indeed they sometimes are—just as administrators and teaching faculty are often put into adversary roles. This is a childish game we sometimes play that the students and faculty—many of whom shift from institution to institution—are going to hire the administrators as a bunch of janitors. I do not think this will work.

CAWS

I specifically said *not* as janitors.

GREENFIELD

I think universities have tended to posture themselves as something divorced from the community and have asked the board of trustees to protect the university from the Yahoos without. The problem is not the trustees' holding on to the status quo against the push for change; it is that universities, like labor unions, have been perceived by some people as what they are not, as necessarily beneficent institutions. Certain people never cross picket lines even though labor unions are racist at home in a lot of cases and colonial abroad. Some faculty members have been in the forefront of a kind of change that has worked for the worst. They have pursued active fundamental far-reaching technological change that has adversely affected people here and around the world. What we have to do now is let the technological mechanism work itself out so that everyone is happy. The point is that the trustee should have a concept of being an outside force checking an institution that frequently does harmful things. That is why the notion of a student-faculty union is appropriate only if we restrict its influence to what happens on campus.

I would suggest there is a two-level construct that applies to this problem. One would want the trusteeship check to look into problems of exploitation on the campus. The notion that the university charges $100 for two people to share a room which is nothing more than a slum is disgraceful. Providing a check through the use of advocacy planning is the kind of thing that the trustee concept ought to take into account. The second is to prevent the university from going outside itself to effect fundamentally inappropriate, undesirable, or immoral change in the community at large, whether it be through supporting Project Camelot, an increase in the power of the military-industrial complex, or destruction of neighborhoods in ghettos. We must ask whether the board of trustees is going to assert a constitutionally valid, effective check on the activities of a university that harm other people.

CAWS

In some other parts of the world what you describe is called a board of visitors.

GREENFIELD

We must discuss with some specificity the notion of what the constituted mechanism is. Obviously the absentee problem is serious. I have been to board of regents meetings at the University of Wisconsin and they are simply a joke. The expertise is all on the side of the administrators. There is no way that the regents can provide an effective check because they do not have the information. The regents are as vulnerable as any welfare mother in a ghetto trying to confront some white-collar planner who has all the facts at his disposal. The result is downright immoral.

MORISON

You are now looking at a university as an agent of social change that, by making an investigation such as Project Camelot, is changing the world in a deliberate fashion.

GREENFIELD

It is more than that, however. I do not have a conspiracy theory, but I do have a performance theory. The same constitutional requirements that would require greater representation would, in my view, effectively prevent, for instance, the dismissal of a professor who spoke out for or against the war in Vietnam. The same constitutional checks that require some boundary of the university influence, in my judgment, would also vastly increase the freedom of student journalists and faculty members to probe actively into conditions of the university without reprisal. This is why I was so annoyed at Jacques Barzun's book. Look at Columbia's activities: cigarette filters, real estate, and IDA. It is hypocrisy for the administration to say when the pressures begin to mount: "We are an institution divorced from greater concepts and it is not our concern."

MORISON

I am interested in how far a mechanism within a university can go in telling its faculty what it can and cannot do. Is this the role of the trustees, or the faculty, or the administration?

CAWS

An academically responsible body ought to exist; I agree with Mr. Greenfield entirely on this point. But it cannot be at the same time the legal embodiment of the university. It has to be something different. Nevertheless, the university has to have a legal embodiment of some kind if it is going to own real estate, have money, be responsible, and so on. Thus, I think we are talking about two separate bodies, and part of the ambiguity of the whole situation is that one body has traditionally been supposed to fulfill both these roles.

HOLTON

I have just had the illuminating experience of being at the University of Rome for eight months. It was in turmoil almost every day I was there—often shut down either by students, by the police, or by the authorities. The slogan on the walls on the first days was "Why do we not hear more about Vietnam in the classrooms?", but by the end of the eight months the students painted the walls with a new slogan: "Let us abridge capitalism." They found themselves moving from certain local issues—why are there no political science professors in Italy except for one or two who seem to think that political science stopped with Machiavelli—and they ended up asking questions about the foundations of their society.

I believe a large proportion of those young people, and some of the faculty, wish to commit themselves to a different ideology than the one that has been ruling trusteeships and the faculty's relationships with their students and with their funding agencies. There is a rather general ideological rebellion against the current models—be they the models from the far left (which certainly are in disrepute) or the models with which we find ourselves in our own situation. If one foresees the constitution in the year 2000, it might very well contain additional provisions: It is the right of every person to have the education that he feels he ought to have to lead a meaningful life, to be a unique individual to a degree that is not now granted in any meaningful way. We are seeing, particularly among the young, a reaching out for different models, a different constitution.

We all seem to agree that some modification is necessary. But that body of wise people who will govern the modified university, who see the future clearly and have the world at heart instead of their own small concerns, will not be solely the members of the faculty; they usually have been selected for other reasons, for narrow self-centered achievement aims. We are all groping for a mix that will educate the different parts, so that we can have the kind of informed dialogue out of which governing policy would arise.

KERR

There are two questions I want to raise. One is a question of fact; the other is a question of policy. First, the question of fact. I was rather surprised when Mr. Waggoner said that there is more autonomy in Latin America and in Europe than in this country. I just do not think that observation is true at all. I think it is rather basic whether or not you think we have more or less autonomy under our system.

Second, we have not got to the question of whether we want autonomy in the United States. There are some people here who are arguing for no autonomy in the university and others who would favor a great deal of autonomy. It seems to me that this question is central with respect to trustees. Have they in the United States been a mechanism to give more autonomy to universities than they had when they were under the control of the church or the state? Is it desirable to have more autonomy, or should we be working toward less autonomy?

It is not only a quantitative question. What kind of autonomy and for what purpose? In Germany, the individual full professor has more autonomy than the American full professor, but I do not think you can say that about the university as a whole. You cannot say that about France nor about Russia nor even about England today. It seems to me that for better or for worse we have more autonomy in our institutions than those in any of the major developed countries of the world. The American system of lay trustees has contributed to this greater autonomy and diversity. One must ask whether it is desirable to have this additional autonomy.

SHOBEN

I find a troublesome drift in the way that we have been considering this general problem of trusteeship and, by extension, the primary social leadership of the university as an institution. Mr. Greenfield has argued for values that, in general, I thoroughly share; yet I worry about the extent to which we may be a little time-bound in giving a somewhat hurried and possibly too strong assent to particular ways of achieving these values. When I think about some of the techniques now being considered for promoting greater internal democracy in universities as well as for generating an enlarged concern among academic people for institutions in the larger society, I wonder if we might not attend with profit to the ghost of that prototype of fanatics, Oliver Cromwell. Never expecting to lose the battle of Shrewsbury, but worried about the possibility that his troops would loot the city after their conquest of it, he rode through his ranks, admonishing his men, "Be mindful in the bowels of Christ that ye might be wrong." In the present groundswell of enthusiasm for universities to become highly serviceable in direct ways to the American community, we may indeed be wrong in a fashion that does real harm to the very groups to which the academy should be responding much more positively and constructively. For instance, we might keep in mind the sheer fact that the official personnel of our colleges and universities differ in some fundamental and distressing dimensions from the great majority of minority-group members. Consequently, although any decent-minded man will applaud our attempts to cope more humanely with what Clark Kerr has called "the ghetto clientele" by more open admissions policies, I hope we can at least think seriously about the possible outcome of merely increasing educational access. Until we learn better how to define and provide bona fide educational experiences for people with whom our institutions are too little acquainted, I am not at all sure we will be serving the black community or the university by a more open admissions arrangement As a matter of fact, that arrangement strikes me as implying a promise that we cannot fill (and that may look as if we have no intention of filling it) until we have worked out proper and creative modifications in our programs, our instructional staffs, and our standards of evaluation to insure that Negro, Puerto

Rican, Mexican-American, and other students, new to our campuses, really have *de-facto* opportunities for self-development and the acquisition of the skills and awarenesses that are most relevant for them. Such changes simply won't be easy to bring off in our institutions, and we do not know what the cost—in intellectual and educational coin as well as in dollars—will be even if we manage somehow to get the changes instituted. I certainly do *not* mean these remarks as a plea for a series of studies that will postpone our meeting what for me is a clear and urgent responsibility. I am certainly pleading that we recognize much more clearly than I think we have so far some of the complexities in issues of this kind. They are pretty harrowing for all their sharp moral imperatives.

GREENFIELD

Did anybody even ask those questions ten years ago?

SHOBEN

I suspect none of us did, but I see little advantage in wallowing in either our own or somebody else's ten-year-old guilt.

GREENFIELD

No, but the point is not guilt so much as responsibility. A pretense that we can remain at the status quo while we decide whether or not to go out from the boundaries of an isolated institution is simply not supportable. The question is whether the universities of this country bear complicity not for indifference, but for active participation in activities that have harmed groups that should not have been harmed by what they were doing.

SHOBEN

Quite right. Again, however, even though my values are yours, the issue of community involvement in university government strikes me as enormously complex. It is extremely difficult to argue for the notion of a high degree of internal institutional autonomy, which has its own calculus of dynamic expansion, and at the same time espouse the thesis that there are certain kinds of common and widely accepted social constraints that can be responded to only selectively. Can we really have institutions that respond only to those community influences that the university itself invites or approves of? If a university subjects itself to the kinds of review procedures that Mr. Greenfield mentions—procedures serving purposes that I heartily endorse—then it puts itself in some degree under the jurisdiction of interests quite external to itself. When you begin to get the representation of these other kinds of interests in reviewing a university's program, then it is going to be very hard to keep out those interests that would be less welcome than those to which we presently feel warmly receptive.

GREENFIELD

Yes, it would. But I am arguing for the necessity of making that effort, not for the ease with which it can be made.

SHOBEN

I am not all that sure that the case for necessity has been convincingly made; but in any event, the argument is a matter neither of ease nor of necessity, but of entailments. If we begin to think seriously about community review as an alternative to conventional concepts of trusteeship, then we must ask what we are going to be letting through our gates that we have not yet considered very carefully. I see little basis in this arrangement, for example, for screening out a variety of right-wing influences or of determining on educational or intellectual grounds those kinds of contradictory interests and conflicts of opinion to which universities can productively react. At the moment, we are generally willing and even eager to hear: "You have no business being involved in classified research. You have no business being such a foul neighbor to a poverty-stricken neighborhood from which you get real estate. You have no business serving the ghetto clientele as badly as you do now." But although these are messages to which our ears are now wide open, they are not the only ones we are going to hear if we adopt the procedural alternatives that are being advocated.

GREENFIELD

I think that is misreading what I said. I do not share the view that we should participate in everything that is left and progressive, but shun everything to the right. Is it not odd that SDS, George Wallace, and many people who are not Wallaceites, but lower-middle-income whites are saying the same things about the influence of foundations and the influence of academia? That does not necessarily mean that one can sit back and say: "Aha. Both extremes are wrong. We have to continue to occupy the middle." Both those groups may have legitimate complaints, and we must listen to both of them. I am arguing not only that we have not listened to black people and radical students, but also that we have not heard the factory workers and the Wallace constituency. They are citizens of this country too, and we may have overridden their legitimate concerns as well.

TOLLETT

Their concerns are really a desire for universities to facilitate their own socialization. Their concerns are primarily vocational: They want to get an education that will equip them to earn a salary that enables them to move to the suburbs. I am not sure that you can expect progressive reform to come from the democratization of the educational process. Indeed, I think an important question that should be raised is just what kind of institution a college or university is. Is it a democratic institution simply because it is in a society that is democratic? We automatically think that the university will be better if it is more representative since ours is a democratic society. Yet no democratic implications flow from the premise that a university is concerned with cognitive rationality, as Mr. Parsons says in his paper in the fall 1968 issue of *The Public Interest.*

ESKOW

I should like to struggle publicly for a minute with my bewilderment, because I can no longer connect the cure with the disease. I cannot trace any of the ills of the modern university to the trustees at all, nor can I see any longer how the multiversity would be any different today if there had never been trustees. If the market mechanism is the protest and the protest is listing the diseases, which of these diseases were created by trustees? Were our prescriptive and elitist policies for admission the doing of the trustees? Is the irrelevant and fragmented curriculum a reflection of the value system of the trustees? It seems to me that the university as it exists today is a projection of the wishes, dreams, and aspirations of the faculty. The trustees, in a sense, had almost nothing to do with the design of the institution. I think we are creating an elaborate mythology that suggests that this institution has become an expression of the value system of the trustees—of the military-industrial complex that they have organized with the aid of captains of industry turned administrators. That is a mask that hides the truth. If we have become knowledge factories, it is not because the trustees have wanted to generate knowledge. If we have lost community, it is not because the trustees do not have a sense of the college as a community. If we have been indifferent to the factory workers and the blacks, it is not because of the trustees.

CAWS

Yet the existence of boards of trustees has had the effect of removing from the faculty the necessity of thinking about some of the questions we have been talking about.

GREENFIELD

You cannot at one point say, "We have to be very careful about going into the ghetto," when you have an active, encouraging, free-wheeling, eager participation by many of the faculty in the damaging activities that now go on. You cannot consult corporations, the Institute for Defense Analysis, send missions to Vietnam, study pacification of insurgencies abroad, and at the same time say you shall discourage or be very prudent about going into the ghettos, about encouraging anti-war protests because, after all, you are an isolated community. You cannot have it both ways.

SILBERMAN

It seems to me that we are blurring two rather separate questions. One is the question of whether the university should be an agent for change or a body for the accumulation of knowledge. The other is the unintended consequences of what we normally think of as the search for new knowledge. We are now discovering that many university undertakings have rather large consequences for change outside the university, and that we must take these into account. Columbia may be buying property on Morningside Heights only to permit its own expansion, but this expansion may have detrimental consequences for the community in which it is located. To talk simply about the university as a place where knowledge is sought and not as a service station ignores the large consequences for society implicit in many of the things a university does. We cannot get away from that question.

Much of the discussion so far does not bear much relationship to my understanding of the history of the American university. As I understand it, the trustees did not take away from the faculty the power to choose the administration. More often it was the other way around. The primary academic freedom at Columbia, certainly, is the freedom from any responsibility for administration. In the past, the Columbia faculty chose deans and department heads. At the turn of the century when the university was expanding, the faculty pounded the table and said: "We are not going to be burdened with these administrative chores. We are scholars. The problem of picking deans and department chairmen is a function of the trustees and not something a real scholar ought to be concerned with." Much of the growth of the administrative apparatus at Columbia has been the result of the refusal of the faculty to have anything to do with administration. Faculty may have changed their mind, but the experience at Berkeley suggests quite the reverse. The faculty response to the first crisis was to insist that student discipline ought to be in the hands of the faculty and not the administration. But the faculty then responded to the Dirty Speech Movement by saying: "Do not ask us scholars to be concerned with things of this sort." I think we at least ought to be candid in recognizing what the historical movement has been.

Also our talk of the autonomy of the university seems to exist in a historical vacuum. From the Middle Ages on, the university has not been autonomous; it has been, as Mr. Kerr suggested, a service station for society. Those who are arguing that the university recover its autonomy and search for scholarship are really asking for a return to the ancient role of the servicing of the clergy, the aristocracy, and the elite rather than for the dirtying of its hands with servicing other groups in the society.

Perhaps the question that is most useful here is one Mr. Riesman posed at the April meeting: "Who are the customers?" The suggestion that the students and faculty ought to make all of the decisions and hire the administrators implies that they are the only customers. Yet if we recognize that the university has rather large effects on the rest of society, it is hard to deny that other groups in the society are customers also. The question then becomes one of balancing conflicting claims. At what point do ethical questions merge with administrative questions? What groups in the society is the university supposed to serve? Autonomy is a will-of-the-wisp in any absolute sense.

MORISON

This question of the customers is clearly an important one. Are the objectives of the university truly internal ones? Is the university autonomous in that sense or is its autonomy a procedural thing in order to meet the needs of the customers? This morning we have reached a consensus that there are customers outside the universities as well as inside for which the university has some responsibility. But how much autonomy does the university have in deciding how to fulfill these obligations? I think of autonomy as meaning that the internal part of the university as an institution has a right to choose its own faculty, defend its faculty against outside criticism, and not bow to outside pressure. This is quite different from what happens in Latin America. The university should also have considerable autonomy in selecting its students and research projects. It probably should have some autonomy in determining how it goes about extension activities since many of our universities were set up with extension as one of their main functions.

KERR

I should like to discuss the idea of autonomy in choosing students. Oxford and Cambridge had autonomy in their choice of students. It took a couple of Royal Commissions to get them to take anybody who was a Methodist, or a Jew, or a woman. In Mississippi and Alabama there was also autonomy in the choice of students.

RITTERBUSH

May I offer a statement that might help to resolve the difference of opinion on autonomy? I should like to propose a hierarchy of governance. At the top and most important is the adoption of institutional objectives drawn from the range of available alternatives. This is really the most important function of governance. Subordinate to it, at a second level, is the setting of intellectual and other goals derived from those objectives. And then, on a lower level still, are questions of strategy relating to the mode of pursuing those goals. It seems to me that Mr. Kerr has been saying that the American university has been autonomous at the highest of these levels, freely altering and adopting objectives, and what Mr. Caws seems to be saying is that the university should be so stable an institutional type that there need be no uncertainty or outside intervention in setting objectives, which would leave only questions about intellectual goals, which could best be left to students and faculty. The degree of autonomy desirable within and for universities seems to me to depend upon which level of choice is being discussed: objectives, policy, or strategies.

CAWS

All I mean by autonomy is self-regulation.

HOLTON

It is not self-regulation if you keep on doing what everybody cheerfully lets you do.

CAWS

But from the point of etymology, autonomy *means* self-regulation. Self-regulation is not inconsistent with being held accountable by some sort of body.

HOLTON

An operational definition of autonomy that makes sense to me is that you can engage in a qualitatively new venture with fairly good expectations that you can carry it out. It may turn out to be a bad venture, but at any rate you can change things rather than keep doing what the matrix already has been set up for you to do. At most colleges, we cannot suddenly give credit for a course that a group of students wants to force into the catalogue. Our lack of autonomy comes from the structure of the faculty and of the student body as it is set up. Harvard cannot (and I believe should not) suddenly turn around and give credit for courses that are in fact being widely attended—1,100 students are attending courses under the auspices of the Kennedy School, but as extra-curricular activities. When there is a group that wishes to do an experiment, then we have a test of whether this group is antonomous, and of the limits of its autonomy. For example, those students are autonomous within extra-curricular activities, but not autonomous within the framework of college credits.

MORISON

The individual is autonomous in a university in that sense. But the university itself has an autonomy in the sense that it does not always have to check its procedures with some external body. You are just hung up on your faculty.

HOLTON

No. I am interested in understanding the mechanism for changing the rules in order to make a new venture possible. Within a small framework we can make some experiments. We can, for example, suddenly have a freshman doing independent study instead of a specific college course. It is not necessary to go through a major re-examination of the university to let this happen. But I think the operational test of autonomy is how far you have to rediscuss the whole purpose of the organization before you can make a new experiment.

MEYERSON

We were talking over coffee about the financing of higher education, and I should like to relate the problem of the financing of higher education to the problem of student rights, responsibilities, and behavior. The adult backlash running through the country is such that I strongly suspect that within the next few years those institutions with a strong financial base will be those institutions whose donors are dead: namely, Harvard and a few other universities or colleges for which past bequests provide both the capital and the income for a splendid or potentially splendid ongoing program. I suspect there is going to be a tremendous reluctance to provide money through the means that we have known heretofore for the support of conventional kinds of collegiate and university programs and, I fear, even less inclination to support unconventional programs. I say this with great sorrow, but I think, for example, that what we see currently in the State of California is not due to Governor Reagan alone. When we find a political leader of great subtlety, such as Jesse Unruh, saying much the same thing, we begin to see a pattern in which funds will be sharply reduced—not absolutely reduced, but reduced in terms of the needs that come both from an increase in scale and from the increasing desire on the part of the faculty, students, and others to have far more flexibility.

Let me proceed to discuss certain possibilities that exist before us. The next few years will see gigantic debates on the financing of higher education. The public institutions have formed a coalition with the aim of getting funds from the federal till. For the most part, the private institutions have concurred in the aims of that coalition. The result will be the greatest educational pablum that the nation has seen. Every institution will have a basis of support quite apart from its competence and quite apart from the imagination of its programs. There is an alternative suggestion that we single out a handful of institutions as national institutions, making sure that they get the most significant kind of support. Such a proposal would freeze into the system a pecking order and would reflect past achievement rather than potential. If one were to single out a dozen institutions, who would say that Columbia should not be on the list? Yet from many points of view, it might still have to prove itself.

All of this leads me to the proposal that Zacharias and others have framed. I think their proposal has the greatest merit. They suggest that it ought properly to be the student who receives the financial where-with-all—the token, if you will—to enable him to go to the institution of his choice. Let me proceed to indicate some of the virtues of that proposal. I can imagine few things that will give the students a greater influence over colleges and universities than their control over the dollar. If a student has through a grant or a loan—and in most cases, especially in the near future, it will be a loan—the funds to pay for his tuition, his living arrangements, and probably some kind of cost-of-education allowance to the institution, the vote of the student would determine the financial viability of many of our institutions. This would not be true of those institutions where the donors are dead, but it would hold for most other institutions.

I also support the efforts to do away with large numbers of required courses—almost a return to Charles Eliot's university—because natural science and other fields will be put on their mettle. A chemistry department may now get five hundred students in beginning chemistry simply because the students have to take a set number of science credits. If the students do not have to take beginning chemistry, you may get a significant intellectual advance in the teaching of chemistry. A chemistry department will be most reluctant to lose five hundred students.

Similarly that institution that cannot appeal to the student who comes with his token for tuition and board will lose him. In some cases, the student will choose the institution that is the least rigorous and the least demanding, but I am far more hopeful that students today instead will choose subtle kinds of educational experiences. They will go to the institutions that are trying to improve themselves and are trying to develop a level of educational excellence far superior to that which we now know. Many institutions will "suffer" if we permit the use of that rhetoric. I think people rather than institutions suffer. Institutions ought not to have human characteristics applied to them. They can come into being, and they can go out of being. If they go out of being, there may well be some good reason why they go out of being. This past Congress passed legislation encouraging universities to punish students who have engaged in disruption. Such legislation will, I think, quickly become standard for all federal aid programs to higher education, and I suspect it will become the standard for legislation on the state level with even more alacrity. State legislatures in all the states are much more subject to the adult backlash than Congress is. One of the safeguards against the results of that reaction is an approach that provides money to students. It is a very minor safeguard, functioning much the same way that the

G.I. Bill functioned a generation ago and has functioned to some extent since then. I mention this last point in passing and tentatively; I am not convinced that such a program of aid to students will have the political character that I have suggested.

The main reason in favor of this proposal is a very simple one; it is a crude form of economic control. Those who in the past have had economic control over our institutions have had many other kinds of control over them as well. Let us not deceive ourselves into thinking that they have not. The controls may have been direct or more likely indirect. It is a rare meeting of the governing board of any state university, for example, that is not constantly concerned with what the legislature will think of its action. It is a rare meeting of the board of trustees of any private or public institution that is not deeply concerned with what the alumni donors will think of the actions of the students, the faculty, or the trustees. I am not disputing the pattern of trusteeship at this time, but I am saying that giving students the money to provide for their own education will produce a greater responsiveness to student needs on the part of the 2,200 colleges and universities in the United States than we have ever had before.

I am not willing to predict what the outcome of this redistribution of resources will be. It might well carry the worst features of camp existence. We might end up with the same kind of collegiate institutions that made Mencken wince in the 1920's. Yet I believe that this process would encourage a set of colleges and universities to be more responsive to a different and important set of educational requirements and to be intellectually far superior to anything we have ever had before.

MAYER

I could not agree more with what Mr. Meyerson has said. But there are certain colleges and universities that will not lose public support to the extent that Mr. Meyerson suggests; these are the colleges or parts of universities that are devoted to technology and to the professions—colleges of agriculture, engineering, nursing, public education, home economics, and so on. For these, support is going to continue to increase because this increase is necessary for the operation of our society. Furthermore, students in colleges other than those of liberal arts, by and large, seem to have had far less difficulty in finding the meaning of their education than have liberal arts students. Yet all of the non-liberal arts colleges as well as some of the graduate schools—medical schools, for example— have given insufficient attention to the arts, the history of their own science and technology, the philosophical and ethical aspects of their profession, and indeed the place of their profession within the over-all context of the nation. Reinforcing the liberal arts content should not necessarily take the form of additional courses. Rather it should be a change in the general atmosphere. This is very much needed, particularly in this time of increased specialization, and the liberal arts college could make a much greater contribution than it now does to those other schools of the university where the public expenditures are not going to decrease and may perhaps even increase.

MEYERSON

We tend to forget that the overwhelming majority of American undergraduates are in professional programs, because most of us are from universities where that is not so. But the great bulk of undergraduates are in business schools, in engineering schools, in teacher training programs, and in a whole series of other vocational programs. Indeed, depending on what definition you choose for narrowly defined professions, as many as two thirds to three quarters of the undergraduates in the country may be in such programs. At many universities the great bulk of the so-called graduate students are in fields that have been designated as professional.

An astonishing change has taken place between the events at Berkeley and Columbia. At the time of the Free Speech Movement at Berkeley, there were no professional students in the Free Speech Movement—almost no engineers and no architects, for example. At Columbia the students in the school of architecture played a critical role within the series of spring events. The involvement of law students in dissent has grown enormously during the past several years throughout the country (and incidentally at far greater risk to themselves than almost any other group of students because it is a rare bar association that will admit a member who has a history of dissent or especially a history of arrest). For the first time you find medical students expressing disdain at their curriculum, their program, the conditions under which they study, and so forth. If somehow we could create the financial mechanism whereby students are granted levels of choices that they have not had before, we will find a tremendously impelling force for the remaking of these curricula. It is madness to assume that a student who is convinced at the age of twelve that he wishes to be a physician must go through a rigid four-year undergraduate program in which the closest experience to being a physician he has is with a test tube in organic chemistry. He has no exposure to the opportunities that such a calling provides. He is so removed from humanistic training in medical school that by the time he has completed his medical degree, his internship, and his necessary residency, he has ceased to be the educated man he may once have been. Even in their admissions policies, medical schools still tend to favor the boys who did well with their test tubes in organic chemistry. But the students are resenting this more and more. A student in our medical school came to see me in June. He had been at Columbia as an undergraduate where he had had a brilliant record. He then went to work in a psychiatric clinic in Harlem for a year. In his first year of medical school he was near the head of his class in half of his program and in the other half he was near the bottom of his class. He was asked to leave the medical school and is now at another university studying history. But he came to see me before he decided to leave medicine. His argument was that there is something terribly wrong with medical education if a student can do well in part of his program and so badly in another. No program should be so rigid that it cannot recognize that the physician cannot be all things. His plea made me feel dreadful. This man could have made a tremendous contribution to medicine. If you give the students opportunities for choice, I suspect that many of the difficulties that we are talking about would alter.

SILBER

Would you favor a system that would differentiate the amount of financial aid made available to students from families with incomes of less than $10,000 from the remuneration provided for students from wealthier backgrounds? Under such a program, how would one protect the educational system from the application of Gresham's Law? Would the colleges not start competing for students by means that would be contrary to the interests of higher education?

MEYERSON

If we expect large-scale financial support of the kind I am talking about, it is far more likely to be on the basis of loans rather than grants. If it is largely on the basis of loans, then I would worry far less about the social origins of the student, because the loans would be repaid through some linkage to future income and earnings. Nevertheless, many problems arise. What happens to the woman in such an arrangement? In Sweden where such a pattern existed for many years, through the Swedish insurance companies, there was a joke that by the time a woman had paid off her university loan she was past the menopause. I am also assuming that although a large part of the income of institutions of higher education would come from the student tokens, a significant part would come from other sources as well. Those other sources will help to provide a leavening agent.

When we follow through on Mr. Silber's Gresham's Law problem, there is much confusion. Suddenly we see de Gaulle possibly being forced to devalue the franc, having just conducted a campaign to force the devaluation of the dollar. Which was the bad money driving out the good? Was it the pound, the mark, the dollar, the franc, or what? It just is not clear. Of course we are all concerned about bad educational money driving out good, but which is the bad money? It may very well be that the student token system will drive out what may now be highly valued currencies. We have got to assume that in the history of university life we must have a certain continuity. How do we make sure that there are certain programs that persist even though tokens are not used for them? This question is, of course, one we constantly face in our own institutions. Obviously classical archaeology these days attracts relatively few people, but it is a rare one of us who proposes the abolition of archaeology departments. We would have to make sure that we do not let the students, through their choices, fully determine the history of higher education.

GREELEY

I am inclined to agree with your recommendation, but I wonder about the mechanics of it. Would you permit the tuition to reach its natural level, or would you subsidize the schools to make up the difference between tuition and cost per student?

MEYERSON

I do think the question is a little more technical than we probably ought to get into. It is a very rare service in our society on which we place a single charge. This includes public services as well as private ones. And yet for some unexplained reason we have assumed within universities that we have got to have common charges. There are people who do not hesitate to pay a surgeon a huge fee, but assume that in universities there ought to be a common denominator—not only among universities, but within a university as well. There would have to be a variety of checks in this financial plan, and there ought to be a vast difference in the amounts of money provided for particular kinds of programs and educational experiences.

METZGER

I agree with Mr. Meyerson that the Zacharias plan probably offers us the best answers. It seems clear to me that across-the-board subventions to institutions would finance mediocrity, that selection of certain institutions for institutional aid would reward past performance or present reputation—both unreliable guides. The Zacharias plan does raise some problems, however. First, it appears that giving the student the right of choice would reduce the residential cost advantages that certain institutions currently enjoy. This might have an adverse effect on community politics, or polities, or institutions. You might say that those that cannot survive are victims of the market place, but that is also an argument for the reduction of diversity in higher education, and whether we want to have institutions that are in every way national would be a question. There is another question—quite the opposite of the Gresham's Law fear—the fear that the good and the bad will be all right, but that the different will go under.

One other issue that I think we must recognize is the possibility that the giver will lay down behavioral and attitudinal tests for the receipt of funds. Is it not possible to penalize an institution that deviates? Is it not possible to penalize the students? Those that have been aware of the pressures of loyalty oaths and disclaimers are aware of this danger. The question of whether it is or is not a certainty depends on the degree of legitimization that the university manages to secure. The question of legitimization and all it implies must be explored. There still would remain a need for categorical aid to finance the expensive research equipment that modern science and scholarship require. I take it that for the Zacharias plan to go into operation current modes of financing and all the problems they involve would automatically disappear.

Then finally I should like to raise the question of whether or not the Zacharias plan would truly equalize access to educational opportunity. I have no doubt that it would go a long way in that direction. That was not a reason you gave for the program, but it surely is an important one for financing higher education in this way. It makes it possible for the poor to get an education. Yet, on the other hand, one must not forget that there is a problem of foregone earnings. Individuals bear much more of the cost of higher education than is represented by their tuition and their residence expenses. They forego the money they do not earn while they are in school, with the exception of their summer earnings. I wonder whether it would be possible to imagine a Zacharias plan that would take that into account; I assume that the costs would be astronomical.

KEEZER

What has been the rate of repayment of the student loans thus far?

As you know there are mixed histories. A number of loan programs have had substantial defaults, but others have been administered in such a way that repayment raises few problems.

In answer to Mr. Metzger's point on the reverse of Gresham's Law, I would count that as one of the benefits. Let me use for examples the case of the two-year colleges and the needs of the Negro colleges. The two-year colleges are disparaged around the country. Some are marvelous places for educational experiences, but in many parts of the country two-year colleges are indistinguishable on the level of instruction from secondary schools. The student continues to take elective courses he did not have a chance to take when he was in secondary school. The value to any student of that kind of experience is slight. Yet in many parts of the country students are forced into these two-year institutions and have no other options open to them. The egalitarian nature of the Zacharias program would enable students who do not presently have other choices to have them. It would force many of the two-year colleges in the country, which are administered by the secondary schools and run like secondary schools, to change their character vastly. Such a change would be all to the good.

In the case of the Negro colleges—and here I would add the church schools, too, since many people go to these schools because they wish certain associations, certain identifications, and certain kinds of moral and other experiences that they would not have elsewhere—I would not in any way suggest that those factors would become less important. A study was made of where the G.I. Bill of Rights students went to school, and we learned that a somewhat higher proportion went to the Catholic and the Negro schools. I am not at all certain that it would mean the demise of these schools if we had a considerable upgrading of all institutions.

The other main point is that of foregone earnings. I am currently running a program for 125 black students who in virtually all cases have to have stipends provided to their families—their own primary family or their parental family. It would be impossible for those families to function without this supplementary help. I have managed to scrounge enough money to finance this program for this year, but I do not know how we can continue it for very long. These problems reflect the whole character of our social structure; they cannot be dealt with through the educational mechanism alone. If we had some kind of compensatory redistribution of income to low-income families, the problem would, at least in part, solve itself. Certainly on the short run it is almost inconceivable that resources would become available for dealing with the foregone earnings problem. We do deal with this in the case of graduate students, but that is the only place in higher education where we do so.

PAYNE

It seems to me that student choice in this kind of program makes a good deal more sense than parental choice, which is what has been going on in most places. The problems of precocity and dependence are likely to be better met under a system where there is a certain amount of freedom from the financial dependence on the home and the scholarship department of the institution. You have a possibility for a more self-confident student body in that way and that is a great advantage. I am worried, on the other hand, about an additional problem of freedom. The mere fact of having to start paying off loans immediately after college is one more way of closing in a student's options. It might be feasible to talk about outright grants plus loans. This would also make it possible for an economic calculation to be made that would lead to some students staying in the community colleges because they would in fact be more economical in the long run.

REVELLE

If we follow through on Mr. Meyerson's idea at least as far as the professional schools, we have to add another element to the entities that are involved in the governance of the university, and that is the professional associations. At the present time the professional associations largely determine the requirements for entrance and practice in a profession. On the whole, their influence is, I think, pernicious in professional education. Unless we can find a mechanism whereby the professional school rather than the professional society has a major voice in determining the requirements of the profession, we are likely to freeze into the Zacharias proposal not the free choice of the students at all, but the peculiar educational notions of the professional associations.

We might also look at where this kind of allocation of resources has been tried, though in a somewhat different way. In the Philippines, certain universities exist as private profit-making corporations. These institutions obviously depend on the free choice of the students, since they can somehow provide what the students want and make money at it. On the whole, they give a poor education from every point of view except the short-run interest of the students. The students receive their union card, a piece of paper that gives them status in the society. The program Mr. Meyerson is advocating should, I think, be highly experimental; it should not be considered *a priori* as the way to proceed. I was glad to see that Mr. Meyerson did comment that we probably need a wide spectrum of kinds of support. Our experience in the relations between the federal government and the universities has been disappointing. The project system for support has come about as close to ruining our universities as any single device could because it has led to a loyalty, particularly on the part of scientists, not to their institution, but to their so-called peer groups in Washington. A man in any of the scientific fields, at least until recently, could move whenever he wanted to. He could take his grant, his graduate students, and his whole apparatus with him. He was, in fact, not a member of an institution at all; he was a member of a kind of national club that cut across institutional lines. If we reason by analogy, it seems to me that any single way of supporting higher education on the part of the federal government is liable to bring unforeseen disadvantages that may turn out to be much greater than the advantages.

SPROULL

There has been a great deal of talk about Gresham's Law here today. I just want to point out that one of the problems that we all face now is that the Washington corruption of Gresham's Law—that the urgent drives out the important—is now applied to universities. Universities are the last institution in society that ought to accept or apply that law.

I do not agree with Mr. Revelle that the Philippine experience is relevant to the proposal Mr. Meyerson has made. Surely that situation reflects the free choice of parents rather than the free choice of students. If it is a profit-making institution, it is surely a place where the parents are paying for the education.

My real point has to do with incentives. We all feel that there is an institutional insensitivity and lack of responsiveness in major American educational institutions. Here I am talking about the six-million-student problem and not about the Harvard or Cal Tech problem. But how can one achieve institutional responsiveness or institutional sensitivity? Mr. Meyerson has, I think, leapfrogged his way over a vast amount of territory in saying: "Let's put incentives into the institutional support structure in such a way that the institution has incentives to be responsive and sensitive." I should like to ask what other things are happening in institutions represented around this table to increase responsiveness and sensitivity in the shorter run.

GRAUBARD:

Perhaps Mr. Metzger might comment now on his experience at Columbia.

METZGER

I am not clear on what level—descriptive or analytic—I should proceed. I assume that you do not want a blow-by-blow account of what happened at Columbia last spring. And I doubt that I can improvise a summary of all the large meanings of those events. Perhaps I would do best to seek a compromise—to tell only so much of the story as will illuminate one of its main motifs.

Let me address myself to the problem of authority—a theme which any storyteller from Columbia is likely to think of first. Last spring, a large number of Columbia students not only violated the rules of the institution, but challenged the legitimacy of the rule-makers. A self-constituted body of the faculty interposed itself between the students and the administration and attempted to mediate their differences. This *ad hoc* intervention failed, for reasons that shed a certain light on why *ad hoc* interventions of this sort often fail. The administration called in the police (not once but twice within a fortnight and each time with bloody results). Failing to anticipate, constrain, or condemn police misconduct, the "authorities" of the institution lost most of the moral credit they had left. For a time, there was no authority at Columbia, but two competing and incompetent seats of power—an officialdom that could summon force, but could not gain the consent of the governed; a student strike committee that could shut things down, but could not gain and keep allegiances, even within its ranks. Gradually, a committee of the faculty, set up by professors assembled in full strength, acquired sufficient moral standing to become, if not a new authority, then at least the instrument through which authority might be restored. It prevailed upon the trustees of the university not to press charges of criminal trespass against students accused of one offense; it supported the efforts of another body to judicialize the discipline of the university; it appointed a commission of outsiders to investigate the causes of student anger; it began the framing of a constitution that would give students and professors a more important role in the running of the institution. Whether these techniques will work—whether a calculated policy of forgiveness, a more thorough commitment to due process, a willingness to submit to outside judgment, and a more democratic form of governance will reknit our shattered community—it is still too soon to tell. The way the spring term ended admonishes against facile optimism: The official graduation ceremony had to retreat to a nearby church in order to protect itself against disruption; a radical student counter-ceremony took over the embattled campus grounds.

I think it can be said that at no other university up to that moment was the attack on authority so jugular, the crisis of legitimacy so complete. Yet, when one looks back, one finds little evidence that this attack and this crisis were anticipated. Those in charge of the university knew that students were restive, but they felt that student restiveness was a chronic illness, not one that reached climatic peaks. A minority of radical students was attacking the alleged seduction of the academy by the social institutions of war and profit, but it was believed that numbers were decisive, and that those who were complaining about the CIA, ROTC, and grades for selective service could command the soap boxes, but not win the troops. I would attribute this false sense of security in part to conceptual failings. What produces and sustains authority? President Kirk seemed to employ a "rational-legal" model: Authority, he appeared to believe, is the gift of the charter and the statutes conferred on the holder of a specific office. Vice-President Truman seemed to use a pluralistic-pragmatic model: Authority, he appeared to believe, is the product of the capacity of a system to recognize and satisfy divergent interests. The senior professors seemed to favor yet another concept: Authority, many of them seemed to believe, is an attribute of personal competence, as attested by institutional credentials and by disciplinary repute. A sanguine corollary could be drawn from each of these conceptual models. President Kirk made the Weberian assumption that as long as he stayed within his sphere of competence and obeyed the general rules, his orders would be obeyed. Vice-President Truman seemed to infer from his political science that as long as students had access to the administration and could negotiate with it for concrete rewards, they would be socialized into the going order, even as the Populists and the CIO had been in another day. The senior professors seemed to think that of all authorities, their own, a product of mandarinism and charisma, was the most natural, the most winning, and the most secure.

What these models failed to take into account was the thrust of black student militancy and the animus of white student revolutionism—two very different forces which joined at Columbia for one frightful week. To the black students on campus, who had gathered in numbers large enough to make the assertion of race consciousness a collective rather than a lonely task, academic authority was white authority and thus congenitally suspect. They did not test the moral quality of an official action by asking whether it was authorized or *ultra vires:* In their eyes, the rules, though general, were not neutral with respect to race. Nor would they automatically accord to a white professor the respect due knowledge and renown, for knowledge could be tainted by prejudice and fame could be accorded evil men. This did not mean that they would never treat with white officials. To the extent that they regarded the university as a dispenser of occupational advantages and the arbiter of racial rewards, they did conform to the Truman doctrine and bargain for specific gains. But black confrontation with white authority was no longer just an ask-and-get affair. It served an expressive function: It was a way of acting out resentments, a way of exorcising inner doubts; it involved a purging of the servile images incorporated into the black man's psyche through centuries of social subordination. To use a convenient shorthand, the black students were concerned not only with de-exploitation, but also with de-victimization. The goal of de-exploitation is a fairer allocation of resources, a more even division of the pie. In the context of the campus, it took the form of demands for greater black enrollments and for larger scholarship funds for the poor. The goal of de-victimization is to repair the psychic consequences of prolonged subjection. Whatever forms it takes, it cannot have so finite an agenda. Historically, the prime vehicle of de-victimization has been de-colonization—the replacement of the foreign superordinate by the indigenous subordinate in every position of power and prestige. The demography of this country may not permit so drastic a solution. Still there is room for colonial analogies: Population movements have concentrated blacks in ghetto pockets; administrators, social workers, and policemen, entering the "native quarters" by day and leaving them by night, do bear a certain resemblance to foreign officers, religious missionaries, and expeditionary armies. The fight for community control of schools is somewhat akin to a bid for sovereignty, although it is more partial and oblique. When Columbia decided to build a gymnasium in the park that separated Harlem from the Heights, it failed to consider these analogic possibilities. On its face, the move was not exploitative. The facility would be used by both the community and the college; located in a city park, it would not force any tenants to be relocated and reduce the limited housing stock. Ten years ago, these considerations had weighed heavily with Harlem's leaders, who had helped pass the enabling laws. But now a new set of leaders, reflecting a new black power mood, was inclined to see things very differently. To them, Columbia was a white settler's city on the hill—a Salisbury, a Johannesburg, a Nairobi—rising high and mighty over the native flats. Like them, it had seized the land around it through a policy of purchase and chicane and had expelled the native population so as to protect its safety and hygiene. Now, its appetite unappeased, it was about to appropriate more tribal territory that had struck it as desirable and green. No matter that it was willing to share the building's floor-space, the control of the building would not be shared; moreover, the very plan for sharing betrayed an *apartheid* prejudice—one entrance on the high ground for students, another for the lesser breeds below. I am not certain that every student at Columbia who was black saw the gym as a colonial intrusion. But almost every student who was black took sides with the black community, rather than with the white-led institution, once these Mau Mau symbols had been raised. To put race above status was to prove that one had not been suborned by white preferment; this was essential to self-esteem, and self-esteem was now of the very essence. (I recognize, as I say this, that Columbia had been harsh toward its neighbors, and that this imagery, though exaggerated, had a germ of truth.)

At the start of the week that shook the ivy, Hamilton Hall was seized and occupied by black militants and white revolutionaries joined together. Soon, however, this alliance fell apart. After a night of squabbling, the whites were sent out into the early morning cold to colonize another building, and the blacks had Hamilton to themselves. Why this separation? In part, because the blacks demanded autonomy. (In a crude way, the building was for them a kind of country; the negotiations for its release, a kind of foreign policy; the removal of the white contingent, an assertion of their own self-determination.) In part, because the blacks did not trust their allies. (At first, the whites opposed the barricading of the building and proposed to block administrative rooms instead, seeking thereby to avoid embroilment with the student body. The blacks interpreted this to mean that these white rich men's sons in their Skid Row clothes were willing to take the risk of words, not deeds. To prove they had the courage of their convictions, the whites jettisoned selective tactics when they struck out on their own, and the idea that confrontation should be discriminating in its choice of target fell a casualty to this contest of zeal.) But as much as anything else, the blacks parted company with the whites because of differences over ideology. Where the whites spurned the materialism of society and regarded social mobility as an anaesthetic, the blacks maintained a strong commitment to the stratification system and to conventional notions of success. Where the whites saw the enemy as "capitalism," the blacks saw the enemy as "racism," and these symbols, while they touched, were not the same. Most of all, they had different attitudes toward academic authority. The black militants wanted to face the "Man," both in order to redress their grievances and to address their wrath.

They sought, therefore, a responsible and responsive adversary. (It is interesting that, of all the student groups in the buildings, only the blacks failed to call for the resignation of Mr. Kirk.) The whites would not render the administration the tribute of being needed. In the revolutionary credo, the social system was congenitally corrupt and those who served to sustain it had to be attacked and overthrown. The weakness of university authority, a sometime source of despair to the black students, was for the whites an opening to be exploited, a point in the social organism where one could inject the redeeming plague. The blacks used ingestive and obstructive images ("let more of us in," "stop the juggernaut"), but they did not use toxic or gangrenous metaphors.

At Columbia, there existed no faculty body that could debate important academic issues, generate a consensus about them, and make sure that that consensus took effect. (There was a University Council, but it was dominated by administrators and traditionally had concerned itself with bagatelles.) Consequently, the brunt of this assault upon authority was borne by the emplaced authorities, which is to say, by two individuals—the president and vice-president of the university. To say that it was not borne well is not to ignore the difficulties under which they worked. They were— to start with—only two, hardly an optimum number for gaining the sort of intelligence and instigating the sort of argument that were needed to correct mistaken views. Men younger than they would have been overtaxed; men less able than they would have been overwhelmed. Then, too, they had inherited from the last great president of the institution, Columbia's own demonic St. Nick, the worst of all possible executive worlds—too much uncommunicativeness on high policy to receive the benefits of feedback, too little control over subdivisions to implement any central plan. In a time of crisis, this combination of an overabundance of prerogative with an insufficiency of power turned out to be particularly perverse. Added to all this were the pressures that outsiders exerted on them. Alumni who had been students in a more tranquil day saw their successors as engaging in a temper tantrum that would grow if it were in any way placated; men of property saw property rights invaded and demanded instant recourse to the law's defense; other administrators begged ours to hold the levee, lest the radical torrent engulf them too (this was a liquified version of the domino theory and doubtless just as hyperbolic). For a while, these pressures were resisted. The separation of the black students from the white students had created an all-black citadel in a predominately white institution that stood cheek-by-jowl with Harlem. To allow them to stay ran one sort of danger, but to dislodge them by force ran another—the danger of arson and riot—and the balance of risks had to be gravely weighed. The hostaging of a dean also bought the sit-ins time and allowed them to spread to other buildings. But procrastination was not adopted by the administration as a policy; to the external publics of the university, this would have been taken as a sign of weakness, and the regard of these external publics counted much at Low. This is not to say that a Fabian approach would have been adopted had there not been so much prompting from outside. I do not know exactly at what point the president and vice-president concluded that negotiations with the SDS-dominated strike committee were useless, that the issues of the gym and of IDA (an institute of weapons analysis with which Columbia had nominal connection) had been raised to embarrass not reform the university, that the longer the sit-ins lasted the more they became a counter-government competing with the legal one for mass support, that so ultimate a challenge to authority had to be met by something no less ultimate—the arbitrament of force. Perhaps they reached this point when the SDS took over the president's suite and began to affirm the revolutionary counter-ethic that establishment property, being theft, is therefore eminently seizable and that privacy is never a right that can be said to adhere in official persons. With the blacks, who were less wounding in words and conduct, the administration did make an effort to come to terms; it would not, however, pay the ransom asked for—the abandonment of the gym; and so these negotiations did not prosper. On the third day, the vice-president told the faculty that he had reached the end of his political string. He was about to submit to the harsh protectorship of the police.

But a group within the faculty would not accept this, and their intervention was to produce another chapter in the crisis of authority at Columbia. I shall not burden you with the details of how the Ad Hoc Faculty Group took form. Let it suffice to say that it arose spontaneously, rather than by any plotter's will. Nor shall I try to examine all the motives of those who joined it. Assuredly these were varied and in certain persons quite complex. It may be enough to say that, as far as its steering committee was concerned (as a former member of it I can describe its mind with modest confidence), there was little sympathy for the insurrection. One fear brought us into action—the fear of violence on the campus—and one hope sustained us through four sleepless nights—the hope that third-party intervention, free from the intransigencies of either side, could find a placable and equitable way out. Subsequent commentators have charged that this was a partial fear and a naïve hope—the Ad Hoc Faculty Group has not received what one would call a glowing press. I do not doubt that there is some substance to these charges. We *were* more concerned about imported violence than about the violence boiling up in our midst (though we did constitute ourselves a constabulary to try to keep the left- and right-wing students from each other's throats). And we *did* misjudge the temper of the revolutionaries, which was far more Sorelian than we anticipated. Between our desire to pacify the campus and their desire to traumatize and radicalize it, between our sense of politics as accommodation and their sense of politics as war, stretched a gap that was probably unbridgeable. Yet, it seems to me, there was more wisdom in our course than has been noticed or than even we at the time realized. When the police came with their covered badges and flailing nightsticks, when they not only made arrests but inflicted punishments—and on the innocent and the lawbreaker alike—when they disobeyed the instruction of their own officials in venting their fear and spite, they did more to jeopardize authority than ever did the students at their barricades. To have tried to sustain the life of the university around the confiscated buildings would probably better have preserved the social compact than to have invoked what turned out to be an indiscriminate and ungoverned force. Moreover, it is possible that, in time, the area of disaffection might have shrunk. By dealing substantively with the issues, the Ad Hoc Faculty Group took cognizance of an important fact—that most of the students in the white communes were not revolutionary and had not written their manifestoes tongue-in-cheek. If most shared with the revolutionaries a radical critique of society, they did not draw the same nihilistic inferences; if they agreed that the university had been corrupted, they did not agree that its improvement was beyond all hope. The very innocence of our rationalism allowed us to discern complexities—to see that the buildings had not been occupied by eight hundred anarchistic Rudds, that there were many communards who were not *enragés* and who were tense under the latter's leadership, and that time and a concessionary stance might result in their defection.

And yet I now see that our intervention was doomed to fail—not because it was inept, but because it was *ad hoc*. To salvage authority, one needs authority—or at least some authorization; to exercise power without anointment, to mediate at no one's behest, is almost to guarantee defeat. We were made aware of this in many ways. In our first phase of mediation, when we simply tried to repair communications between the students and the administration, we were seen as abdicating our role as teachers and turning ourselves into colorless connecting links; in the second phase, when we tried to impose a settlement and speak with our own authentic voice, we were seen as making promises we lacked the mandate to put forward and the power and stature to enforce. By going our self-made way, we became an object of suspicion to colleagues who valued due procedures. Some seemed to think of us as intruders, well-intentioned perhaps but too presumptuous; while others seemed to think of us as adventurers capitalizing on crisis to advance our own careers. Perhaps the worst of the liabilities of our *ad-hoc*ism lay in the indefiniteness of its contours. Who belonged to the Ad Hoc Faculty Group? To what extent did the Faculty Group speak for the faculty as a whole? The debate over the issue of amnesty made it clear that these questions were critical and that the answers would be indistinct. "We will not accept judgment of punishment from an illegitimate authority," said Mark Rudd, referring to the corporate authority that emanated from the legal fount. The *ad hoc* steering committee, which was largely made up of senior faculty, proposed that the disciplinary punishment be light (for those who voluntarily left the buildings), but that some sort of disciplinary judgment be made—in short that the students be granted amnesty *de facto*, not *de jure*. Critical as they were of the administration, they did not regard its authority as illegitimate; much as they sought a settlement, they would not agree that the revolutionaries were in the right. Doubtless, for professors who were not so critical of the administration or were not so eager to arrange a pact (and these included some of the most distinguished members of the faculty), even amnesty *de facto* went too far toward immunizing lawlessness and acknowledging its moral claims. But in the group that called itself the *ad hoc* membership, there was a surprising amount of "faculty" sentiment in favor of a prescriptive amnesty. A certain part of this variance in sentiment was clearly an artifact of composition. Existing outside the rules, the Ad Hoc Faculty Group had no standard by which it might define itself except that of an incontinent egalitarianism. Some who came to vote and argue had fewer ligatures to the faculty than to the students (some indeed *were* students, with relatively minor teaching responsibilities). Some were of a rank and office that would have barred them from a seat on the established faculties or would have admitted them without a vote. In regular academic governance, there may be something to be said for open doors and an elastic franchise; in irregular modes of operation, there is not. Because we did not know what we were, we could not maintain confidentiality; whatever happened in our sessions was instantly reported by double agents to the SDS. As the votes on the issue of amnesty teetered back and forth, depending on the vagaries of attendance, the student rebels came more and more to believe that "the faculty" might declare itself their moral ally.

Clear-minded observers would have known that, whatever happened in these frenzied meetings, the bulk of the tenured faculty would not dissolve the thousand ties—personal, contractual, and customary—that bound them to the given order, and the idea that they would have preferred to live in anarchic confraternity with the students rather than in orderly relations with the legal powers should have strained credulity from the start. But crisis is not a time for clear-minded observation, and the rebelling students did believe that they would get large pickings if they stood pat.

The Ad Hoc Faculty Group failed. The lessons of its failure have not always been correctly drawn. Its failure does not prove that whenever authority shows cracks it crumbles. "United we stand and divided we fall" is a precept for preserving power, not legitimacy; a flawless alliance between Columbia's professors and administration might have brought the police in sooner, but would have made for greater difficulties when they left. Nor does it prove that mediation in a polar world is fruitless. True, our evenhanded formulas were rejected—politely by the administration, contemptuously by the students; we did seem, from each perspective, to be disingenuously partial to the other side. But one rejection after five days' trying can hardly be generalized into a law of nature, and it remains to be seen what mediation can accomplish when time is less at a premium, and there is less need for a climatic "yes" or "no." For me, the primary lesson of that failure is that structures should anticipate, not trail, events. For what we were trying to do, we needed an established mechanism that could draw on the legitimacies of tradition, that could enjoy collegial proxies, that could define itself. We needed a faculty organ; all that we could manufacture, on the spur of the moment, was a crowd.

The coda to the story I am telling involved the effort to create just such an instrument in the dark period following the "bust." On the day after the police retook the buildings, all the professors from the varied divisions of the university met together in the chapel, one of the few places free of the strains and debris of war. Only once before had they met as a collectivity; this had been on the eve of battle, when the administration had convoked them for the purpose of gaining their general support. On this occasion, the members of the Ad Hoc Faculty Group had been able for the first time to address their peers on the dangers of a stiff-backed policy and the virtues of mediation; they had not, however, submitted their proposals to a vote, fearing that a negative or closely-won result would hamper their negotiations with the students. (Whether they had thus let slip an opportunity to gain majority support for their own position, it is difficult to tell. Clearly, they had made a greater impact on their audience than their pariah instincts had supposed they would. Nevertheless, the only formal outcome of that meeting was a bland vote of confidence in the administration.) Now, in its second convocation, the assembly had to face the consequences of decisive action: almost two hundred students bloodied, four or five thousand students on strike, an armed patrol at every gateway, every civility effaced. The inclination of this body was now to find some way to go back—to retrieve a community based on shared assumptions that had so suddenly developed into a regime of force. It sensed it could not do this by rallying again to the administration; it needed its own instrumentality and one that would not be factional or outcast. Out of this sense was born the Executive Committee of the Faculty, something new under the academic sun. It was given a nostalgic mandate—"to return the university to its educational tasks." But the past could not be recreated; the very existence of this body was a rebuke to the old regime; the disaffection of the mass of students had to be coped with, not wished away. In a short time, the Committee became an advocate of reform as a key to recovery, and "restructuring" became its motto and major aim.

For all its advantages of birth, the Executive Committee of the Faculty had to struggle for legitimation. At the beginning, it did not transcend the deep divisions of the faculty. Its parliamentary sponsors, hoping to have it do so, had fixed its membership in a slate composed of old *ad-hoc*ers, staunch supporters of the administration and a few members heretofore disengaged. This *politique* approach, instead of creating a committee of the whole, created a committee of warring parts, and it took time before the conciliations of an able chairman and the unifying effects of the committee process could work to solidify the group. At the beginning, it was distrusted by the students. In an effort to restore the peace, which it saw as a precondition to reform, it became deeply involved in crisis management and lost credibility as a force for change. Moreover, its notion of the changes called for—a greater voice for faculty and students in a government that would leave old forms intact—aroused a great deal of student skepticism in quarters where "co-option" was a fighting word. It was not until the summer brought composure that a *rapprochement* of sorts could take effect. A radical, but not revolutionary student group, which had broken away from the diehard leadership, turned its energies to "restructuring" and became the critic and the partner of a faculty similarly engaged. With the Kirk administration, the Executive Committee of the Faculty had very strained relations. This was a third block to legitimacy. But for the recognition accorded it by the trustees, the Committee might have found itself in the role of its *ad hoc* predecessor. As it happened, the exigencies of the moment impelled the board to turn, not to their deputies, but to an agency of the faculty for advice. It was a marriage of convenience, not without its points of discord. The board thought the way to restore authority was to invoke full disciplinary sanctions; the faculty argued—in part, successfully—that a punitive university would be a university beset by enemies, that retribution without mercy would lead to confrontation without end. The board wanted to continue to allow the president to raise intramural punishments at his own discretion; the faculty argued—more successfully—that this was as judicially untenable as it was politically unwise. Not all the arguments were won by the faculty and not all the arguments of the faculty were sound. What mattered as much as the score on issues were the utilities gained by the debate. The trustees came to know their campus; the Executive Committee of the Faculty received the accolade of their regard. I think I have located one of the latent functions of trustees—they can validate new agents of authority when the old no longer work.

The returns are not yet in. The relegitimation of authority has not yet been accomplished at Columbia. We are only at the point where we may think that we have learned something from our distress.

DIAMANT

I have a few points at issue with Mr. Metzger on the question of legitimacy. The overt issue at Columbia was the gymnasium, and the question of legitimacy was not whether or not the gymnasium was a benefit to Harlem, but whether the Harlem community was in control of that decision. It was at this juncture that the students disagreed with the university. Mr. Metzger also remarked that *ad hoc* situations are dangerous, especially continuous *ad hoc* situations, and that we must have structures ready to deal with crisis. If I heard him correctly, he said that you do not know who you are when you are *ad hoc*. The comment upset me dreadfully because I do not think that structures can define who we are. Perhaps Mr. Metzger does not think so either since everything he was saying, while overtly calling for structure, seemed to be saying implicitly that part of the problem is that under the structure that exists we do not know who we are. Though he said that his group felt they could not give *de jure* legitimacy and *de jure* amnesty because that would have recognized the illegitimacy of the structure, what he really meant was that the group knew it was an illegitimate power structure. Their real concern with structure and with avoiding *ad hoc* situations arose from a desperate fear of having to fall back on what all human beings have to fall back on—the legitimacy inside ourselves and the legitimacy of our values, our purposes, and our goals. The defense of the administration—which the faculty was being pushed toward much against their will—meant a retreat to an identity defined by structure. In other words, they were saying: "I am a professor; I fulfill my human responsibilities partly through teaching and research" and not "I am a man." There is a tremendous difference between those things. I heard implicit in Mr. Metzger's dilemma the kind of honest intuition on his part that the power structure at Columbia was illegitimate.

SILBER

The thing that I find most perplexing about the Columbia situation is that it came so many years after the trouble at California. After what happened at California in 1964 and at Wisconsin and elsewhere soon after, most faculties and administrations immediately recognized the need to revise their rules. I find incredible the revelation, made at this conference, that there was no faculty senate at Columbia. I had taken for granted that no faculty of any standing could fail to assert its voice, however ineffectually, in a faculty senate.

At the University of Texas, we called for the development of a new set of rules in the spring of 1965. On the advice of a first-rate constitutional lawyer on our faculty, we sought a system that would guarantee all aspects of due process of law and provide the basis for court review for any decision made by the dean of students. The courts could then take jurisdiction in any case in which students were clever and determined enough to insist on court review. We assumed that they would be.

A new period then began. A committee composed of faculty and four students with alternates—the students chose their own representatives, including the president of the student body—set to work. All of the students attended regularly for three weeks, but this was a laborious job. We met two afternoons every week throughout the academic year 1966-67 and through the following summer, a season when it is generally difficult to keep faculty members in town. One of the two students who remained to the end was an alternate. The president of the student body attended less than 20 per cent of the time. The original draft amounted to something like three hundred pages of material, which we thought could be reduced by half. The university administration hired a legislative bill writer to put the deliberations of the committee into the best legal form so that the intent was clear and the wording avoided, for instance, the passive construction that specifies actions without identifying agents.

The finished document was submitted first to the students for their review and their approval. It was then submitted to the Faculty Council. Then began some of the toughest debates that our faculty has ever had. We had to persuade the old guard, who balked at any rule that would give students the guarantees of due process of law and would deprive the dean of students of some of his discretionary power. After a series of test votes on the less controversial points, we decided that we finally had the support of the majority. Then we started passing the sections one after another. It took most of the fall semester to pass the entire legislative program, at the rate of a chapter or two per session. By the time the 1967-68 faculty meetings were over, our set of rules had been passed.

In addition, our president summoned a group to discuss what we would do if disruption took place on campus. We decided that it would be a great mistake to put ourselves in the position of ordering the police on campus. Instead, we incorporated in the rules a statement on disruption on university property, and agreed that if the rules were deliberately transgressed, we would seek a court order under provisions of the laws of the State of Texas. If a judge found that there was a violation of state or municipal law, he would issue a court order that would be presented to the students involved. They would be told that if they did not obey the order they would be in contempt of court. At that point it would be up to the judge to decide whether he wanted to enforce his own court order; the decision would not be left to the university. We knew perfectly well that calling in the police would be a last resort, to be used only if the disruption became so serious that the civil authorities decided that action must be taken. The only time the Supreme Court of the United States failed to support Martin Luther King occurred, I believe, when he defied a court order—not when he defied an ordinance or a rule established by a bus company or a city. But when he refused to obey a court order, the Supreme Court found him in contempt on the ground that there is something ultimately legitimate about the legal system itself. So we placed our ultimate confidence in the legitimacy of the legal authority. (In doing so, I do not think that we were breaking any new ground or being particularly creative.)

These procedures were well worked out, and the students were clearly informed of them so that there could not be any uncertainty about what was permitted and what was prohibited. We wanted them to understand the nature of the legitimacy that they might bring into question: First they could test the legitimacy of the court order; only later could they come to a test of the legitimacy of the university's administration and rules.

The regents have subsequently altered some of the rules passed by the students and faculty, and the faculty has not yet been fully informed of the extent of the alterations. We are trying to have the changes clarified. Since the regents do not meet with the faculty, they rarely have any idea of the extent to which such matters have been thought through. That the regents should change these rules is a source of dismay to many members of the faculty. I think that if the regents had met with us to discuss the rules they would probably have accepted them as they were written.

We took the attitude that the right of revolution is the right of all students, but that the corollary of the right of revolution is that revolutionaries had better win. If the corollary of the right of revolution were the right of amnesty, we would have an open invitation to revolutionary activity. The only thing that protects society from the casual exercise of the right of revolution is the clear recognition by revolutionaries that most revolutions can be put down.

I should like to ask Mr. Metzger how we can account for Columbia's not having responded to what happened in California and Wisconsin. Why did Columbia not have a set of rules that guaranteed students the rights of due process? How do you account for the absence of a faculty senate at Columbia? This absence seems to me to make the central issue at Columbia one not of legitimacy, but of competence. I should hate to think that the incompetence of a president could call into question the legitimacy of the government of the United States. How then did Kirk's actions call into question the legitimacy of Columbia's administration?

MEYERSON

I want to reinforce Mr. Silber's first concern and his concluding one, but also to dissociate myself from the central part of what he dealt with. On the central part, let me be somewhat facetious and refer to an aphorism of Robert E. Lee's when he was president of Washington College in Virginia. He stated to his student body that any boy is worth more than any rule.

I find myself horrified by three hundred pages being winnowed down to half that, but I should like to go back to what I take to be Mr. Silber's concern: How odd that Columbia had not learned. As he talked, I thought back to my own freshman year at Columbia when a late and distinguished professor of classics engaged me in an argument on tragedy. He took me back by saying that the astonishing thing to him was that, having been told by the seer that he was doomed to marry his mother and kill his father, Oedipus proceeded to kill the first man he met and marry the first woman he met. I think what happened at Columbia and at other institutions is close to that.

The kind of homeostatic state we have in universities today leads to a great deal of discontent on the part of students and faculty who want something more Dionysian. The same students who come to me in wrath saying I have taken away their issues are essentially saying that they want a Columbia whether there are worrisome issues or not. A sizable number of faculty, including those who demand that the "weak-kneed administration" bring the police in, essentially wants the turmoil and the aphrodisiac effect that a Berkeley or a Columbia provides. There are those who crave the drama and the histrionics of the moment of disruption.

HURST

Determining the constituents of legitimacy involves juggling a number of factors. The experience of about 175 years of imperfect, but reasonably workable legitimacy in our legal order seems to me to indicate four constituents that are involved in some of the examples we have talked about. A legitimate social order involves the successful assertion of a legitimate monopoly of violence in one place. It is a contradiction in terms to say that you can have legitimate diverse sources of violence. Second, this legitimacy embodies the notion that we call constitutionalism in our tradition. It is basically the idea that a legitimate social order is one in which there is no center of power (defining power as actual ability to control the wills of other people) where the definition of the goals and the means to the power lies solely in the hands of the immediate power-holders. This notion is somewhat paradoxical, but a society must have organized enough diversity so that there is never any single unchecked source of power. Third, and closely related to this, is the value that is challenged by your Dionysian students, faculty, administrators, and trustees. I think our experience with social and legal legitimacy includes the idea that there is a sense in which observance of procedure is in itself a matter of substance. It is naïve at best and dishonest at worst not to see that there are substantive values in procedure and in the way in which you proceed to your value allocations. We have certain notions about what legitimate procedure is. It should be based on rational search for evidence, on an effort to apply reason to the evaluation of evidence. Evaluation, however, should be conducted with a decent deference to the dignity of the individual human being. Fourth, along with the legitimate monopoly of violence which we put in the hands of the policeman, there is the power to tax and spend. This power is a symbol of the idea that one cannot maintain a legal order by repression; it is maintained by affirmative structuring of the situations to desirable ends. Legitimacy is quickly imperiled if power is held and exercised only in negative ways. The only long-run way to maintain legitimacy is by positive responses to demands of the society.

Such a dramatic example as the Columbia situation must not blind us to the necessity of looking at other situations in normal contexts, because the legitimacy of reactions against the system has to be weighed in terms of the context of the system being reacted against. At Wisconsin we have an imperfect system, but it does provide means for affirmative response to demands. The critical fighting issue at Madison in the last couple of years has been essentially the challenge of some of our students and faculty to the substantive value of procedure. I think there has been a great deal of either naïve or dishonest belief that one must seize a building to raise the question of whether the university should hold Chase Manhattan stock because there are no procedures for getting an examination of the university's investment policy. This seems both naïve and dishonest to me because when these people first made this demand, they took over the building. In this context the situation at Wisconsin has been somewhat easier to handle than the Columbia one because if the issue comes down to a refusal to abide by imperfect but workable and rational procedures for value definition and decision, you have a concealed form of totalitarianism. We do have to be prepared at some point to assert the legitimate monopoly of violence, recognizing that this calls for great sophistication and skillful tactics because of the difficulties of controlling the police.

PAYNE

The whole discussion of legitimacy and of rules loses the interest of students rather fast, particularly when what is involved is the protection of the rear flank of the administration. Surely students have an interest in the fairness of the rules. They cannot be expected to participate in the rule-making process simply to keep business going as usual. Students do take the responsibilities of citizenship too lightly. But I am not upset that students fell away from the task Mr. Silber outlined. We are not meeting either the problem of what used to be called student apathy (and is still very much with us) or the problem of the student reaction to desperate boredom—the Dionysian impulses that Mr. Meyerson mentioned. It does not make much sense to talk about due process to people who want change. We need new forms of governance because we need new kinds of things. You have to recognize that students are interested in creation or destruction—not in any kind of stasis.

Higher Education in
Industrial Societies

Very few of those who attended the first session of the Conference on Higher Education in Industrial Societies were personally acquainted with many of those who had been invited. It seemed important, therefore, to have some agenda prepared for the group so that the discussion might proceed more easily. A very brief agenda, prepared by Professors Talcott Parsons and Stephen R. Graubard, was circulated. It read:

Might we begin by reflecting on certain of the more important differences that presently exist between systems of higher education in the major industrial societies? These differences are explicable, in part, by historical circumstances, but also, in part, by the various current situations in the respective societies. It would be a mistake to imagine that the differences will soon be rendered insignificant. Nevertheless, all these societies are today confronted by a common set of problems, and there is increasing pressure to deal with these situations with some reference to what others are doing. Universities, less isolated than they once were, participate in both a national and an international culture.

If our first purpose is to consider how they differ, our more urgent task is to think of how they are presently faced with and responding to common problems. First, how are they reacting to the pressures imposed by numbers? Higher education, until recently, was thought to exist principally for an intellectual elite that was often also a social or political elite. Does the democratization of these institutions imperil traditions and curricula that were never instituted to serve such large numbers? Can these traditions and curricula be adapted to the needs of the new student populations? Will new kinds of institutions have to be invented to accommodate many of those who seek admission? Will new kinds of elite institutions have to be founded? How should a system be differentiated by types of institutions? What is going to happen to the old "professoriat"?

Indeed, what are the variations and changes already present? What are the implications of these changes for the professions? Ought we to anticipate a substantial transformation in the way that men and women are educated for academic life? How will the effort be accomplished to perpetuate competence among those who teach in universities? Is there likely to be increased pressure for a division between teaching and research roles? What are the principal hazards as well as the advantages of such a division? How can institutions of higher learning accede to the demands of society to educate a larger proportion of the young and at the same time maintain tolerably high levels of competence in research and training fields that are not intimately related to the education of these vast numbers?

Closely related to the issue of numbers is the larger issue of social demand. In a situation where public expenditure for higher education is certain to grow rapidly, the question of how universities or research organizations organize and maintain their independence of public authority will become more complex. Increasingly, the assumption may well develop that universities ought to "serve" the community that supports them, and not only by offering a modified apprenticeship system to the young who are being prepared for professional life. The demands on these institutions will grow not only from those who study and teach, but also from those outside who may be expected to assert a more proprietary interest in the institutions they are being taxed or otherwise pressed to support. What kinds of autonomy will universities and other such institutions be able to maintain in this situation? Are we witnessing only the beginnings of vast state-financed systems of higher education? What advantages may be expected to accrue from this new large involvement of the public in the affairs of the university? What disadvantages, if any, may be anticipated? How will the climate of opinion within universities be altered by this development? Will there be a new emphasis given to the "useful" studies, whether they be in the sciences or elsewhere, and a de-emphasis of what may be regarded as luxuries —for example, the "pure" sciences, the arts, and the humanities?

Universities have long been engaged in the training of practitioners in the principal applied professions, from the clergy through law and medicine to engineering. What is the future of this function, and how does it fit with research training and the general education of the more enlightened citizenry?

The dynamism of the intellectual world to some extent casts universities and their personnel in the role of agents of change and critics of any given status quo. Can we generalize about this role with reference to change? What are its implications for the relations of universities to the rest of society, especially government? Must they face new pressures toward intellectual as well as political conformism? But, equally, must they learn to use their enhanced influence more responsibly, and how?

Finally, are new kinds of structures developing that will make the last third of the twentieth century as significant a time in the history of institutions of higher learning as the last third of the nineteenth century proved to be? What are these structures likely to be? How will they affect the more traditional kinds? What "institutional inventions" are most urgently called for now? How can we imagine seeing them accomplished?

Participants
Joseph Ben-David
François Bourricaud
Asa Briggs
Mauro Cappelletti
Michel Crozier
A. D. Dunton
Baron John Scott Fulton
Pierre Grappin
Stephen R. Graubard
Walther Killy
Yoichi Maeda
Martin Meyerson
Elting E. Morison
Victor G. Onushkin
Talcott Parsons
Alessandro Pizzorno
Marshall Robinson
Edward Shils
Sir Solly Zuckerman

GRAUBARD

In the fall of 1964, in what I now think of as the pastoral period of university existence, when all was calm, and if there were clouds on the horizon, they did not appear to be storm clouds, *Dædalus* published a volume called "The Contemporary University: U.S.A." Recently I had the good or bad judgment to reread that issue. The one question uppermost in my mind was who among us saw what the likely concerns of university men would be in the year 1968. Some of those who wrote for that issue were more sanguine than they would wish to be today. But in my preface to this issue, I began with the following statement:

> In a time of rapid change, when almost all institutions assume new forms or show evidences of strain in seeking to accommodate themselves to unprecedented situations, there will always be a certain amount of disagreement about what in fact is taking place. The profound transformations that have occurred in American universities during the last two decades are very reasonably the subject of discussion and controversy. While certain of these changes give promise of unparalleled opportunities for research and inquiry, making possible service to society on a scale which would have been inconceivable before the Second World War, it is not always clear what "price" is being paid for these advantages, or what their ultimate institutional and personal effects will be. It is generally assumed that most of these innovations serve to improve American higher education. The evidences of curriculum reform, new sources of financial support, and the multiplication of libraries, classrooms, and laboratories are certainly impressive; the results of this vast expansion of effort must one day soon show themselves in the lives of those who gain from the new opportunities offered.

In rereading the last sentence, I wondered whether I believed that they *must* or whether I was actually saying that they ought to. All of us are aware of the vast explosion in university populations. With this vast explosion, certain university systems, the American among them, have imagined that they could prepare for the initial demands of these new persons by a proper construction of new buildings—mostly classrooms, dormitories, libraries, and the like.

In the period before Berkeley, many universities in this country felt proud precisely because they had not been caught short in providing facilities. But it is perfectly obvious today that the problems of university education are no less serious despite those facilities. In other words, in a country like Italy the facilities do not exist. One is simply taking the old vessel and throwing numbers of new people into it, making modest curricula reforms with almost no real recognition of the new kinds of people who are coming into the universities. There one feels, and quite properly, that this essential provision for the new university population was never made. In the United States and in the United Kingdom, that provision was made; there were new buildings. And yet we all sense how acute is the problem in all of these societies—both those that have provided the physical facilities and those that have not.

It might be profitable to begin our discussion by talking about whether any of our societies have been able to provide for these vast new numbers. Have we, for example, given enough attention to what is implied when you take essentially elite institutions and attempt to make them popular institutions by simply saying that they can accommodate so many more than they had accommodated before? What does it mean when instead of a few institutions that take the bulk of those who are preparing for the positions that universities prepare for, you suddenly have a vast proliferation of these institutions? What happens in the end to the profession of teaching? Alfred Knopf, the publisher, remembers that before the end of World War I, you could bring together in a single room every historian who imagined that he was doing something and therefore had some interest in what someone else might be doing. As a result of the vast influx of students and the unparalleled expansion of university faculties, anything that we may once have believed to be true in the way of colleagueship, in the way of relations between students and faculty, is clearly in question today.

Thus, I should like to see us begin by discussing this matter of numbers. How much are we all beset by the problem of numbers? How well are any of us coping with numbers? Are we simply accommodating new and larger numbers, or are we indeed establishing new kinds of institutions where these numbers can be educated and feel some security about the kind of education that they are receiving? And most importantly, how are those who provide the education and do the research responding to this vast expansion of the university world? This issue seems to us in the United States to be critical; perhaps it is not so important abroad. Thus we might perhaps start with the problem of the democratization and popularization of our universities. This change has occurred within the adult lifetimes of people in this room.

PARSONS

Not only have the numbers increased in this country, but the people coming to institutions of higher education are from very different social backgrounds than was the case only a generation or so ago. Perhaps that is less conspicuous in the United States, because we have had a system of social stratification somewhat different from that of Europe.

ZUCKERMAN

May I just add a rider to what Mr. Graubard said, since he singled out the United Kingdom as being the exception? The United Kingdom did in fact try "to control" up to a certain moment in time. In practice this ceased to be possible following the governmental decision to implement the Robbins Report of 1963. Up to that moment, we had tried to assess what the rate of expansion of higher education should be in relation to the potential demand for graduates. Today we are, I think, suffering from the same kinds of problems that are being experienced in other countries; and controls are mainly financial.

BRIGGS

From what looked to be simple mathematical projections of likely future demands for university graduates, we passed in Britain in 1959 to looking at the supply of people with what were thought to be the right university qualifications from school. Then from about 1964 we started more or less gearing the number of university places to the output of people from schools who had what were held to be the minimal entrance requirements to get into universities. Given this kind of background, I believe that it is important to talk about numbers before one starts talking about social composition; and I would have a lot of *caveats* to make before I assume that the most important aspect of numbers is social. While there are certainly important changes going on in the social composition of universities—changes that may affect their mood and style—I consider that many of the changes taking place in the expanding universities work irrespectively of changes in social composition. They may, in fact, be the result of changes in the attitudes of the people who come from precisely the same groups who formed the university population before. I think one must test very carefully the assumption that the new student from the working class or the lower-middle class is responsible for troubles of universities.

PARSONS

I did not mean to imply that change in social composition is the only or even the most important consequence of the vast increase in numbers; I merely wanted to point out that it is one.

BRIGGS

I entirely agree that we ought to talk about it, but at a later point. The first issue would seem to be whether in the expansion of systems of higher education you deal with the problem of numbers simply by multiplying students within roughly the same number of units, whether you introduce quite new units, or whether you develop new institutions that are not university institutions. According to the answers that you give, a different resource allocation is involved, and you will get different kinds of universities. We have had examples in every country both of an increase in size of existing units and of new university units being brought into existence, and we should start talking about numbers within the limits of these mathematical exercises. Social composition can then be considered as we get involved in the qualitative aspects of university expansion.

CROZIER

Having faced the same kind of problem, I would like to come back to the point of social composition. Some of the trouble that has developed in French universities starts in institutions where the social composition of the people is upper-middle class. We should think about the social origin of people, but we also have to think about the place people will find in life and the way universities are preparing people for a particular social class. People at Nanterre had expectations that were not at all fulfilled. The students from upper-class origins went to the university with the expectation of a kind of elite arrangement and found something that is no longer elite at all. They were terribly frustrated, and much more so than the students from lower-class backgrounds. I would agree with Professor Briggs. I suspect social origin in itself is not so important. In France, as in England, we have not called in many people of working-class origin or from the lower income levels.

MEYERSON

I should like to return to Sir Solly Zuckerman's point on a self-regulating system that presumably existed in the United Kingdom before the Robbins Report. At the risk of being brash, I should like to suggest that so many of the ills that the United Kingdom suffers from are directly due to that presumed self-regulation of educational demand and supply. The spread of education has been, in very large measure, the key element in the economic advance of the Soviet Union, Japan, and the United States. Tremendous numbers of people have had a set of experiences that have not always led to elegant intellectual results, but they have led to two terribly important things: either to the acquisition of a whole set of scientific and related knowledge that would enable major economic advance to take place or to a change in class position and class expectation.

Half a century ago, you did have an elitist situation. This did not mean that only those who came from an elite origin were in universities, but that those who went to universities could have in the future certain elite roles in society. We are now reaching a point—the United States is probably closer to that point than any other country—in which half or more of the people will expect to have some kind of higher learning. I am suggesting that this model of mass higher education is very appropriate for a country that wishes to have a sophisticated level of economic development. One of the ironies is that so many of the protesting students are saying that they do not wish to have that kind of sophisticated economic development and would as soon dispense with it. This view is, of course, their option; it may indeed even become a societal option. Until it becomes a societal option, however, I would suggest that advanced education is the major means for transformation within the economic sphere.

The United Kingdom achieved its initial industrial advances at a time when the populace generally was not particularly well educated, and the universities utterly scorned the life of trade, the life of the economy. Today, however, economic development does demand an educational level that only universities can provide.

BEN-DAVID

Everybody assumes, because there have been troubles at certain universities, that there must be something wrong with the universities. I would put forward the suggestion that perhaps there is an ecological component here. The site of the problem may be the university, but its cause may not be. For instance, if I take the situation in the United States right now, more than 40 per cent of the age group between eighteen and twenty-two are at the universities. As a matter of fact, they are conscripted into the university; if they ever want to get ahead in life, they cannot do anything else. Thus, the population that is potentially the most active politically is at the university. They have plenty of time. They live away from their parents, usually, and have the fewest obligations and personal responsibilities. So if one had to ask oneself where a social revolution would take place in such a society, one would have to answer the university. Any trouble that gives rise to social agitation will take place at universities. Places like Berkeley and Columbia in this century are the parallel of Paris in the middle of the last century, when young people from all over France who wanted to seek a new future congregated there. If they did not find society to their liking, they made trouble.

According to a survey of a representative sample of Berkeley students conducted in November, 1964, the following distribution of answers was obtained to the statement:

Although some people don't think so, the president of this university and the chancellor are really trying very hard to provide top-quality educational experience for students here.

Forty-four per cent strongly agreed, 48 per cent gave moderate agreement, and only 5 per cent disagreed. Among militants, the percentage of those who disagreed was 12. To a question on how well they were satisfied with "courses, examinations, professors, etc.," one fifth replied that they were very satisfied and another three fifths said they were satisfied. Only 17 per cent expressed any degree of dissatisfaction. Thirty-two per cent of those who were very satisfied, 29 per cent of those who were merely satisfied, 33 per cent of those dissatisfied, and 40 per cent of those two in a hundred who were very dissatisfied were militants. [Cf. Robert H. Somers, "The Mainspring of the Rebellion: A Survey of Berkeley Students in November 1964" in Lipset and Wolin (eds), *The Berkeley Student Revolt* (Garden City, N.Y., 1965), pp. 549-50.]

To a large extent this process is being repeated in the United States. I would not generalize this to Europe. All attempts to use the student disturbances at American universities for the purpose of constituting basic reforms at the university have failed, and we now have four years of experience. As soon as the troubles started at Berkeley, a number of disaffected staff immediately adopted the F.S.M. and pointed out the ills of the university according to their own beliefs. But according to the surveys conducted during the early stages of the movement, even the activists showed great satisfaction with the administration, the quality of teaching, and their teachers. Their subsequent apathy toward the various educational experiments is further evidence that this was not an issue.

There is no concrete sign that there is something basically wrong with the present structure of the American university, except perhaps this fact of conscription. One ought to rethink the absolutely unfounded custom that every person has to receive a formal education, a diploma, and that all have to go through this experience at the same age.

The problems in the American university reside in parts of the university and not in the whole of the university. The university's function has changed for a certain proportion of the students; even ten or twenty years ago, much of college life in the United States was very different from what it is today, especially at the elite universities. Then many people went to Princeton or Yale for the same reason that they used to go to Oxford or Cambridge: to have a good time. In the course of their stay, they also received some education, but it was not terribly important what education. This was an extension of adolescence and a kind of socialization too. Now the universities and the colleges attached to universities have completely lost this concept. They are now creating kinds of professions and an underground college culture that are potentially deviant and disruptive because many young people do not find anything that they want in a university.

Other students perhaps should not have gone to the university because they are too mature for it. I think we can recognize the type of person who is now among the university activists. These are small, marginal problems if compared to the university as a whole, but they can be disrupting and we have to face them. One should not generalize and start thinking that there is something basically wrong with the universities.

I also want to point out the problem of staff. If any generalization can be made about student disturbances here, it is this: A serious outbreak occurred in cases where a considerable number of staff not only condoned but actually implemented the eruptions. These two groups are not fighting the same war; they have quite different objectives and quite different ideas and ideals.

MEYERSON

I should like to ask Professor Ben-David why Israel is almost the single exception throughout the industrial world of a nation in which the university students have not exhibited the dissidence that we know so well from Tokyo, from Warsaw, from Prague, from the London School of Economics, from Columbia.

BEN-DAVID

Unfortunately the answer to this question is very simple. We might have the same problem as the European universities have because, from the point of view of our educational policies, we are somewhere in the Channel between France and England. Universities in Israel have expanded in much the same way as the European universities, without giving much thought to the proper facilities or to the quality of the education. We have not had dissidence simply because our students have a more serious war to fight.

ZUCKERMAN

May I refer back to Mr. Meyerson's comments? I agree with him fully that advanced university education is a prerequisite for economic advance. But other conditions are just as necessary; not one of them by itself is sufficient. The question that still comes to my mind is whether or not we necessarily need the large battalions of university graduates for economic advance. Are the bigger numbers qualitatively the same as the smaller numbers of yesterday? Mr. Graubard mentioned in his opening remarks that there was a day when one could get all the meaningful historians into a room this size. Once upon a time I knew everybody in England who was doing endocrinology; indeed, I helped start the Endocrinological Society in our country and its journal, as well as the first society in England for the study of animal behavior and its journal. I am not certain that knowledge is now being advanced at an increased rate proportional to the increase in the numbers of scientists at work. Nor, indeed, do I believe that more significant questions are being raised.

I was a member of the Barlow Committee in England, set up after World War II to consider the expansion of the universities and the need for more trained manpower. We came to a very quick conclusion; the war had shown that there were not enough scientifically and technologically trained people to deal with the technical problems that had been generated by the war and the phase of postwar reconstruction. Our recommendation was that the intake of the universities should be doubled as rapidly as possible. The committee was not a committee of scientists, and we had to make certain obeisances to old tradition, and one of them was that if the universities did increase in size, they should not necessarily be distorted from the pattern that they had enjoyed up until that moment; that there should also be an intake into all other faculties equivalent to the intake into the science faculties. Needless to say, as you well know, the intake into the other faculties was greater than the increased intake into the science and engineering faculties. I recall a friend of mine in the Treasury, who had to help persuade the Chancellor of the day that the money should be forthcoming for this expansion, warning me that as the universities expanded, we were not going to increase the proportion of first classes coming out of the universities on anything like the scale that one would expect from the increase in total numbers. Nor did we.

One problem in England may not apply to other countries. An enormous number of our science and engineering graduates has been absorbed into an aerospace industry which had grown greatly as a result of the demand brought about by the war. It is still responsible for a big intake of trained people, and this has not necessarily been to the economic benefit of the country. I mention the aerospace industry; another is the nuclear industry. At the present moment, we are spending about fifty million pounds a year on nuclear R and D of one kind or another, using the best people we can in a field where there is enormous competition and so far little payoff.

Another result of the great increase in the output of university graduates is the great increase in the volume of so-called research that is pursued. I was pleased to see that a vice-chancellor in England has recently commented that far too much mediocre research is now being done in many universities.

So I return to the basic issue: Is it not possible to conceive of some better controls than those that have been discarded? I accept that the diffuse market mechanism is no way of handling the problem. Professor Ben-David has implied that some of us who were at universities in the thirties and twenties went there because we wanted to have a good time. In those days the British universities had an extremely distinguished record in scholarship and turned out many who helped advance the economy. Even so, I would reject the older market mechanism.

After the Barlow Committee reported, I became the chairman of another committee, whose work was subsequently dismissed in something like two sentences in the Robbins Report. We had tried for every bit of nine years to devise techniques to see whether or not one could get some prevision of what the rate of expansion should be in relation to the social and economic development of the country. Most were inadequate, and the last one did not have a chance to be tried. In effect, the Robbins Committee said all such attempts are hopeless. Their report went on to say let us now throw the universities open. With what results? I am not talking now about student unrest, but of a period of very cold weather for the universities due to a lack of financial resources. The United Kingdom economy cannot sustain the rate of university growth that is now demanded, any more than it was able to sustain the rate of growth of our Atomic Energy Authority, which at one moment ran at 15 per cent. Physical facilities are falling off; staff remains a difficulty.

But what about the practical effects of the vast expansion of the universities on the economic state of the country? I have looked at some new figures that show what the immediate trends are. At this moment in time, the rate at which university-trained people are heading into industry is increasing fast, and has been going up over the past four years faster than anybody could conceive. Yet the performance of the British economy still leaves much to be desired. I, therefore, should like to know whether or not it is necessary to consider just one kind of higher education when one considers economic advance.

May I end with a tangential observation? A fortnight ago I was written to by a distinguished headmaster in England, asking my advice about a sixteen-year-old boy in his school who happens to be a brilliant mathematician. He has already taken an external degree at the University of London and got a first. The following question was put to me: Should he go to a university and possibly be destroyed trying to broaden his education? I wrote back saying be careful about broadening his education. In a century there may be four mathematical minds which can penetrate new veils and transform knowledge. Don't broaden his education at the moment: wait and find out whether the environment of a university would not blunt his genius.

I therefore return the question to Mr. Meyerson: Must we really educate the whole public as a necessary condition for economic advance? Does that mean that the doors of a university must be wide open?

MEYERSON

I thought it would be unbecoming not to defend the American national honor earlier this morning; thus, I made a case for an egalitarian mass-based higher learning. If we do not use a market mechanism or a kind of surrogate of the market mechanism, what mechanism are we to use in allocating resources for universities? Certain patterns are evolving throughout the world whereby increasingly the resources for higher learning are coming through government funds in a combination of national and state or provincial sources. Those allocations are being made not by the market, but by certain presuppositions as to what a university ought properly to be doing and what its priorities ought to be. The independent latitude of the University Grants Committee in Britain is altering considerably. The counterparts of the U.G.C. in the rest of the Commonwealth are probably becoming even more rigid. The Australian grants committee has formulae as to what ought to be, and Mr. Dunton knows the provincial formulae that are used north of the United States, formulae that build in certain sets of expectations. With the exception of only a handful of institutions, American universities, whether public or private, are now dependent on public aid. With each year that goes by, there will be more dependence on public aid. This public aid will increasingly have certain expectations built in—namely, the more students, the better; the more scientific and technical work, the better; the more students from across class lines, the better.

MORISON

If you begin with Professor Ben-David's proposition that not everybody who is at a university is appropriately there at this moment, you can begin to think about ways of dealing with the numbers problem. There are probably two parts to that question. First, what is the appropriate structure of the university if it is to be a distributor of talent? It may be overloaded now in trying to obtain too many different kinds of talent and disperse them too widely into appropriate places in the society. One has to ask oneself what the shape and character and purpose of universities ought to be today, and then one has to ask if there are many students who within this definition are not appropriately there. Where ought they to go? What appropriate alternatives are there to university training? In many instances such alternatives would require the invention of new kinds of institutions. You have overloaded the universities because there is nowhere else to go.

BRIGGS

Before we ask questions about the future, we must begin with the system as it existed before new strains and pressures were put upon it. If we are talking about whether there should be other kinds of institutions to deal with certain categories of people, then we have got to take into account the position, say, of the *grandes écoles* in France. We have got experience of alternative ways of handling groups of students in the past. Before we get to the fascinating point of the social pathology of university students at the present time, we have got to clear away the resource allocation pattern in straightforward quantitative terms. We have to face the marked contrast between, for example, the British way of handling numbers and the Japanese way of handling numbers.

After all, the notion of a university *system* is a relatively recent one. Different constellations of institutions have been emerging with uncertain goals. The first bit of sorting out that is needed is more study of the components of systems, before we get to critical points about determination of future policy. We should ask questions about whether the embryonic systems that are emerging are satisfying the needs and the demands placed upon them at the present time.

PIZZORNO

We have been talking so far too much of a university system as such and not about university relations with the society. We are looking for causes within university systems and do not see that everything probably comes from outside in a way. We should try to see if the position of the university in the society has changed very much and, more specifically, if the relationship between the university system and the professional complex has changed. I think that the real causes of the strains on university systems are changes that have taken place in the professional complex. As Sir Solly pointed out, even in a period of economic stagnation, certain industries exert more and more demand on manpower. In the thirties in Europe, there was a low ratio of degree-holders to total population and, at the same time, there was unemployment among degree-holders. Now we have an increase in the economic growth rate, and yet there is almost no unemployment among degree-holders. On the contrary, we have an increasing demand for degree-holder manpower. Somehow, the whole economic system has been distorted. You can do almost anything in a modern society, but only with very great difficulty can you keep down the output of the university. This is really the only point that the whole society agrees to accept as some sort of a constant. If we look at what happens to the ratio of administrative employees in industry, in the whole society as such, we see this continuous increase as a very abnormal phenomenon. In a way there is an artificial relationship between university and society. Much of student questioning of societal goals comes from their awareness of the artificiality of of the equilibrium between society and university.

DUNTON

Mr. Chairman, I think to some extent we are talking about different things. Certainly Sir Solly is looking at something else; perhaps that is partly a reflection of the different philosophy of higher education in Britain. As in many things, Canada is between the United States and Britain so far as opportunities for numbers of people are concerned. Since World War II, the primary function of universities in Britain has been to turn out highly-trained people. The typical honors graduate of an English university is presumably capable, if he has a first or second class, of going on to professional work in that field. Even if he goes into industry, he should be capable of operating as a professional. The sociologists tell us that we are now moving into a post-industrial era. I suggest that universities always should provide opportunities for people to become highly skilled as research scientists or as historians or as classical scholars. But we should also be allowing young people to leave universities not as specialists, but with a strong ability to adapt. We deliberately ought not to train for some existing piece of research. We ought to try to sharpen their minds so that they can go on learning and do many jobs that will not necessarily be in industrial production, but in all kinds of services and activities. I suggest that one of the big questions before us is what educational opportunities in the universities should be, apart from the obvious one of turning out the people who go on to Ph.D.'s.

ROBINSON

Asa Briggs clarified the problem by stressing the idea of system. We are told from time to time in this country that California is an indication of what the United States as a whole will be like in about twenty-five years. So far as I can tell the system of higher education in California has emerged to a rather high degree in response to folk values. There has been relatively little planning, until fairly recently. But there has been a conviction that education somehow was good and should be provided for virtually everyone who manages to show that he has the capacity to operate in that kind of environment. The environment is not well defined, but someone on the outside can clearly see that there is a well-established pecking order; the natives, however, seem to go on blithely innocent of the elitist elements in the system.

To a considerable extent, it seems to me that this characteristic runs through higher education in the United States. Somehow we have convinced the citizens of Mississippi or Nebraska or Utah that they should be happy with the University of Mississippi, Nebraska, or Utah, and they send their children to it, and Harvard goes on and does its distinctive work. We have allocated certain of the most important functions of the universities to a very few universities. For the most part, these few are private universities. Indeed, Berkeley, the greatest exception to the private university-excellence correlation in the United States, was almost a historical accident, which is rapidly disappearing under the tender ministrations of the processes that are making California more like Nebraska. We have in general been willing to turn over a high proportion of the quest for knowledge to a handful of institutions. A certain amount of the purposeful quest for knowledge, which is sought by governments and corporations, we allocate to another group of institutions. The rest are large multi-purpose higher education institutions, where the central function of the faculty is simply to keep up with what is going on so that they can do an adequate and responsible job of teaching.

Such a system requires an arrangement in which the academics are not very close to the social decision-making system. They cannot be in a position to determine what should happen to whom and indeed what the system itself should be like. It is my impression that what passes for educational planning in this society has for the most part been done by people who are not learned, though many of them have gone to institutions where those wares are made available. They are not manpower planners, though some working suggestion that education and economic development may be important is surely a part of what goes into the thinking on the subject. Because academics have not played an important part in educational planning, we have escaped for the most part the propensity to make each new unit a replica of that which the academic knows. Most men's educational philosophy is autobiographical.

MEYERSON

I have always regarded California as a developing country. You have in California a history of university life and of higher education more generally that reminds me of Edward Shils' monograph on the intellectual in India. The kinds of roles that historically have developed have been very different at different points in time. At the end of the nineteenth century, Henry George, the social critic, could write that the University of California was a place where the taxes of the poor went to support the education of the sons of the rich. The shift that has taken place there has been an extraordinary one. There was a rather self-conscious view on the part of the political leaders of California that education was the second great resource of the state, the first being its climate. There is a conscious sense that certain kinds of consequences will follow if education becomes a major effort in public investment. Those claims are constantly made. Even the people who oppose the University of California at Berkeley are still very much in favor of other levels of the higher learning in California. They point to the aerospace industry in California and a whole series of economic achievements that they believe would not have taken place without huge public investment in higher education.

The notion that the pecking order of the status system of California higher education is not perceived by the members of that system is, I think, inaccurate. The students as well as the staff in the universities, four-year colleges, and two-year institutions have clear understandings of their status. You enter the university if you are in the top eighth of your class; you enter the state colleges if you are in the top three eighths of your class; and you enter the two-year colleges if you are in the residual category. If you are at a university, you can have graduate students or at least doctoral students; if you are at a four-year college, you may have a few graduate students, but no doctoral students. Thus, there are status distinctions that possibly can be hurtful.

I should modify Marshall Robinson on another point as well.

There is a fair number of public universities today in the United States that are probably equal in the level of their scholarship to all but four or five of the private universities, but those public universities may suffer a decline in the period ahead. State legislatures may become less disposed to provide funds for what they regard as erring students and erring faculty.

I think we have been dealing with global phenomena in our discussions, rather than with what happens within the components of universities. I should like to use the illustration of medicine. Most American professors of medicine and physicians will look with great pride at the Flexner report of half a century ago. This report pointed out the dismal character of American medical education, the lack of facilities, the lack of a scientific base for medical instruction. The report concluded that there had to be a tremendous diminution of the number of medical schools; indeed, the author determined that all medical schools were properly to be at universities. As a result of this report, there was a reduction in the number of medical students. The Johns Hopkins University developed the first of the great American medical schools, and it became a model for all American medical schools. It is a rare university in this country that has less than about one quarter of its budget going to the medical school within the university.

The kind of planning that took place was planning by the guild of medicine, and it was a national kind of allocation essentially in the hands of the physicians. They are, of course, thoroughly delighted with the outcome, but I should like to raise in contrast engineering, which in the United States, unlike some other countries, is located far less in the technical institutes and much more within the universities. At the time of the Flexner report, one could have argued that engineering and medicine had remarkable similarities: Universities had schools of engineering and medicine in great profusion; the resources for each were modest; students could with great ease get into either kind of school; and the scientific base for each was of a most casual character. Since then, medicine has become more and more demanding of resources and seemingly the quality has gone higher and higher. I say seemingly because one of the social consequences of the planned constraints on medicine has been an extraordinary shortage of physicians in the United States. This shortage is so severe that it is a rare city in the United States in which the hospitals could function if they did not have their medical interns and residents from Europe and from Latin America and Asia as well. In all but a handful of hospitals, very many of the medical people are now from overseas. Thus we have built up an exquisite kind of selection, as a result of which the rural areas go without medical care and hospitals depend greatly on the brain drain from abroad.

In engineering there has not been this central planning by the guild. The universities that had engineering faculties half a century ago continue to have them. If a student wishes to study engineering in the United States, he is almost certain to find an institution that will accept him. What are the social consequences of the use of that market model? During the half century you have had a sharp differentiation among the schools, so that the best engineering schools have developed more and more of a scientific base for themselves. The most capable people have come typically from a handful of engineering schools, but a tremendous number of people who fill necessary roles within the social system of the United States have gone to the lesser schools. They make up the critical middle group without which American technology could not function; in medicine there is no middle group. Indeed, the general practitioner is disappearing in the United States, so that we have the anecdote about the woman who goes to the physician with a cold and he asks which nostril is disturbing her so he can decide which of his colleagues she should see.

I could have used the case of law, rather than of engineering. The law clerks or assistants to the Justices in the Supreme Court of the United States essentially come from, at most, half a dozen of the country's law schools. There are occasional exceptions, but they are rare. But you also have endless other opportunities for legal study.

I make this case for the virtues of a kind of market mechanism of allocating places and resources as against some kind of central decision-making mechanism primarily to stimulate discussion.

PARSONS

In the medical case, should one not also mention the Rockefeller Foundation?

MEYERSON

Yes, it would not have been possible to have had the Johns Hopkins Medical School without the Rockefeller Foundation.

ZUCKERMAN

The Flexner report also had an impact in the United Kingdom. But the situation that led to the Flexner report was the feeling that there were medical practitioners who were not qualified to practice medicine. There were people around who were putting the public at risk and something had to be done.

MEYERSON

That was the case. Nevertheless, the consequence of the report was the creation of a domestic corps of magnificently qualified physicians who became more and more specialized and turned increasingly to research rather than to practice, with the result that we are essentially staffing our hospitals with the very kind of people that the Flexner report was so concerned that we eliminate.

ZUCKERMAN

Flexner might have added a rider to his report saying that while we want to see that all medical practitioners are properly qualified, there ought to be a second tier of *feldschers*. We have precisely the same problem in England; if it were not for the people trained in medicine from Pakistan, from India, and from other countries, we too would not be able to run our hospitals.

MEYERSON

This issue relates to the social-class problem that Michel Crozier and others have mentioned. What should one call this second tier of engineers? Engineers or draftsmen? Given the social egalitarianism in the United States, it would be most difficult to get anyone in the second tier if it did not have professional status. Many people called engineers are performing a kind of *feldscher* role and yet we give them professional status. Similarly in medicine, we would have the greatest difficulty in creating a subsidiary group of paramedical personnel of high quality in the United States if we did not give them professional status.

BRIGGS

I think we should look at the three possible ways of extension implicit in the notion of increasing systemization. The first way is increasing the size of the individual university units within the system. Here there has been a marked divergence between the European or American experience and the British experience. There have never been very large universities in Britain; there was no great pressure on the universities that were already reasonably large to become very large when the question of expansion was mooted ten years ago. They have all grown, however, and as they have grown, they have undergone organizational strains and some changes of pattern in the process of growth. But they have not been subjected to the complete strains which have so affected European universities that, as several people have claimed, they may destroy the conception of the traditional university as it has been inherited from the past. In all systems there has been much growth concentrated within some of the existing units, but there are differences within the different systems.

The second way is creating new university units. I myself would want to draw a sharp distinction, however, between the creation of totally new university units and the upgrading of institutions that have previously had some kind of subuniversity status. Here again there are differences of approach among countries and different university systems. In some cases, new universities are being brought into existence without any system of tutelage. This was true of my own. Sussex was created because it was decided that the total number of university places in England was inadequate and that new universities were needed. Once it had actually been created, however, it was free to move along the lines that it wished; in fact, its own planning was, to a large extent, autonomous planning. It was limited planning in certain respects, but the fundamental approach to the curriculum and the organization of the university were freely determined. It is important to look into this question of the upgrading of old institutions and the making of new ones with attention to the degrees of freedom that are open to new universities.

Thirdly, we cannot forget the point raised this morning by a number of people. This concerns the third way—the creation of new institutions that are not universities in the ordinary sense of of the word. In all our countries there are institutions serving specialized, limited purposes. In England there is a whole constellation of institutions—ranging from colleges of education at one end of the scale to technological institutions of various kinds with subuniversity status at the other. These institutions fall into the higher education system as it is evolving, but do not have what is called university status. Clearly there are differences in status in different countries. In some countries, the structure allows for more or less the same names being applied to the qualifications given by these different bodies; in other countries, you have quite sharp differentiations. In all countries, I should guess that the sharpness of the qualifications has been diminishing somewhat in recent years; certainly in England this is the case.

If you look at these three ways to expand a system, bear in mind that your adaptability depends on the way you start off. If you look at the expansion process in this way, then you are dealing with questions with "ought" in them, questions about the new kind of institutions you *ought* to found, about adaptability and variation within structures.

Another set of questions concerns the disposition of resources among the different units. Clearly, in certain cases the existing units have been expanded so sharply that resources were inadequate to make the expansion pattern possible without friction. If there had been no student revolt of any kind, there still would have been financial and administrative problems arising out of growth.

Yet we have to look not only at the relationship between expansion and resources, but at the relationship between these institutions and social need. Can we, in fact, produce certain kinds of new institutions that are subordinate to or different from universities, but which will meet certain categories of need more effectively and more quickly than universities might? This problem has been posed in the United States and certainly in England, where there has been talk of the development of a completely separate non-university sector of higher education, including polytechnics and other institutions. In theory such institutions would be able to react more quickly to assessments of national need than the unisities could. I have never seen any very satisfactory figures in relation to Britain about relative costs of development along those lines. Without cost figures here, it seems to me the practical policy questions are not easy to answer.

CROZIER

I should like to comment on the policy from a slightly different point of view. Asa Briggs has emphasized the tendency of all the different institutions of higher learning to become part of a system. Higher education was integrated into a system in France much earlier than it was elsewhere. I fear it may be extremely dangerous to overemphasize the necessity to integrate. If you do, you run into problems of a different kind related to the possibility of having responsive institutions. This issue raises the difficulty not of seeing the system as a whole, but of seeing each institution as one whole system which is answering particular needs. From a certain point of view one may consider wasteful certain activities that from a a longer view would not appear to be so. Mr. Meyerson's remarks on medical education and the Flexner report are relevant here. In many different countries, once you have professions, they tend to use their monopoly over a certain kind of knowledge to their own advantage. Once you have an integrated system, it is terribly difficult to prevent monopolies from developing. So I would advocate some type of sophisticated planning to develop a market mechanism to offset the guild monopoly.

BRIGGS

I was not, myself, doing anything more than describing the problem. I am a great advocate of freedom for diversity within the system, and I am in complete agreement with Professor Crozier. For example, if a new university is tied too completely to the practices of existing institutions, it will not release the creative power open at the start of a new institution. I would, therefore, want to look closely at the conditions within which a new institution is purported to exist so that I could be sure that it would have the power to innovate. It is important for a university not to have too much of a blueprint. It is also important for it to have access to more than one source of finance. The vice-chancellor of such an institution must be an entrepreneur as well as an organizer. The danger about upgrading old institutions is that they often start with an inferiority complex and, therefore, a certain lack of creative power is built into them. From the start, new institutions of a non-university kind have been thought of as being in some sense inferior institutions. The idea that we can get around all our problems by creating a completely new genus of institutions is too simple an answer. We are striving for a system that enables a community to dispose of its resources efficiently, but at the same time allows for creative initiative within the component parts of the system.

CROZIER

The efforts to rebuild the university system in France have run into terrible trouble due to established patterns of patronage. Each new institution is an offshoot of an older institution. Most of the planning has been only a discussion about who is entitled to do what. These institutions have been completely lost as active innovators.

MEYERSON

I am fascinated that we have not yet had any discussion of an international system of higher education. In the medieval world we would have assumed that the world of learning was indeed one that crossed national boundaries. Today, when the world is much closer not only in terms of time, but in other important respects as well, we have little sense of an international university community, of an international system of universities. The medieval student normally would have expected to be at more than one institution in the course of his education; that is not the expectation today. Indeed, in some of the countries represented around this table, it is most difficult for someone who is not a citizen of that country to hold a chair. France would be a good example of that.

My next point is of a very different order. We must distinguish between system as a descriptive concept and as a prescriptive one. My own field of study is that of cities, and we do deal with systems of cities. We could describe the system of cities in each of our countries, and in each case there would be a strong reflection of a series of other patterns within our countries. It is no accident that in France the system of cities has a hierarchy that is similar to the university hierarchy with a tremendous preponderance on the primate city of Paris. But that approach is different from discussing a prescriptive system of cities. I can point to a developing country in the world that concluded there had to be a rank order of cities, and because it was missing certain cities within that rank order, it had to create them. The result was a failure in resource allocation, because there is no cosmic mandate that there must be such a sequence of cities any more than there is a cosmic mandate that there must be a certain sequence of universities and other institutions of higher learning.

Nevertheless, we are moving more and more to prescriptive national systems, because the national state is increasingly becoming the key source of support for universities. It is essential that we try to consider the allocative principles that ought best to be used. The great difficulty in trying to establish such an end system is that we tend to think of most public goods as being instrumental. If we invest so much in transportation, it will have certain tangible benefits. In higher learning we have much less sense of what these benefits are. The tendency, again, is to attempt to put them in instrumental terms. Perhaps universities can contribute to economic development or to certain other purposes, but what we must achieve is a set of allocative principles based much more on the intrinsic characteristics of education, rather than on the instrumental role that a university can play.

We now recognize that the central authority will enter more and more into the resource pattern of university life. How can we make sure that that central authority operates in such a way that we avoid a rigid system and provide an opportunity for wide autonomy? How, in Michel Crozier's terms, do we stimulate a kind of market mechanism by using public aid? A number of us in this country have sharply contended that the consumer of higher education has the best sense of what is intrinsically as well as instrumentally valuable in a university. Let the student have the vote by assessing him whatever charges there may be. If he cannot pay those charges, see that he gets sufficient funds through loans or grants or a combination. There are many dangers in that kind of quasi-market mechanism, but I should like to suggest that there are far greater safeguards in that kind of policy than exist in other methods of central decision-making.

BEN-DAVID

One can envision a Sherman Act for higher education whereby any university that had a monopoly position in the market would be broken down and share the market with another university. I believe that the Department of Justice starts to interfere when any firm has more than 20 per cent of the market in a certain product. There is no single institution of higher education in the United States that has a monopoly position in this sense. Neither Harvard nor any of the elite institutions produces 20 per cent of a given type of product, as would be true of the University of Tokyo or Oxbridge in relation to the higher civil service.

SHILS

I want to comment on the Sherman Antitrust Act and its application to universities. The Sherman Antitrust Act was intended to curb the powerful. The proposal Professor Ben-David is making would curb the inferior institutions. It is not the leading universities that have the monopoly of the output of postgraduate students. The proportion of the output of Ph.D.'s from the leading universities is shrinking relative to the total output of Ph.D.'s in the country. In quantitative terms the center of development is moving away from the major universities. The biggest departments, with the largest numbers of Ph.D.'s, are not always the most distinguished. Would one want to break up the most distinguished departments because they are producing people of better quality? And what difference would it make to break up the less distinguished departments, many of whose graduates do useful research and perform beneficial services for society?

The question of the differentiation of the higher educational system raises the question of the flow of influence among the universities, and this raises questions of autonomy. The question of the permissible and desirable degrees of autonomy of universities has come up in our discussions repeatedly. That seems to me to be one of the central questions that we face here. As long as there was a system of units that were not highly competitive with one another, many places slumbered indifferently of one another's accomplishments and were satisfied with what they were doing. There was a hierarchy of canon; but a position lower down did not injure the self-esteem of those in the lower ranks. They were therefore immune to the pressure of the model of the greatest, and in that sense they were independent. There were centers of initiative and of creation that provided the models which then diffused to other universities. Now, because of increased dependence on central sources of finance, a new kind of system has emerged in which the different units form a whole. Of course, it can never be a totally unified system. In France until recently there was more or less a single system, but even there it was not completely a unitary system. Complete self-government or autonomy of universities has not always been a wholly satisfactory thing under all conditions; on the other hand, the complete central control that existed in France was not satisfactory either, and it has finally broken down. In reaction to the French system, everyone is execrating central control. Yet the complete freedom of a university from any sort of control or pressure from the outside might also lead to a state of slumber. The freedom of Oxford and Cambridge up until the middle of the nineteenth century was complete, except for restrictions on admitting dissenters for degree programs. What kinds of autonomy ought to be safeguarded in the growth of the national system? I agree completely with what Asa Briggs said about the need for multiplicity of sources of financial support; nonetheless, the main source will, I think, increasingly be the central government, at least as far as this country is concerned. For other countries, the central government is practically the entire source.

PARSONS

We have been talking about the institutions other than the full or standard university, and I hope this theme will not be lost sight of. The British technical colleges are one type with a relatively definite vocational focus; the American junior colleges are quite a different type. Both of these stand in the academic pecking order well below the university. In France the phenomenon is reversed; the *grandes écoles* seem to have preempted prestige away from the university. Michel Crozier's article in *The Public Interest* [Fall, 1968] points out that the universities in France are below certain other institutions in the prestige order, and that this might have been one factor in the genesis of unrest at the university level.

A somewhat similar problem arises with reference to the organizational segregation of a city. Powerful forces are operating in the American scene to pull research into the university. The Rockefeller ex-Institute/now-University case is a striking example; it does not want to remain segregated. The Soviet Academy of Sciences enjoys a special position; it is not organizationally part of the university system. This pattern is repeated in a number of other countries, with modifications of course. A colleague of mine said casually not long ago at a lunch table that with all these Columbia and Berkeley situations, there is going to be a major movement in the United States to set up separate research institutes where people are not going to be bothered with unrest. This prediction raises interesting questions as to whether or not the delicate balance between being separated and privileged and yet responsive to demands and subject to certain controls is at all stable in any of our systems. If it is not a stable balance, which way are changes likely to be going?

CAPPELLETTI

Degrees of difference emerge from country to country, but it is fairly clear that there is a differentiation of superior-inferior institutions. The technical schools arise out of the need for vocational, professional, practical training. You will train people in a more practical way in such schools, but they are not universities. At the same time, these proliferating institutions tend to become universities.

PARSONS

That tendency is very strong in the United States.

CAPPELLETTI

My second point is on research. I have spoken recently with friends at N.I.H. [National Institutes of Health] in Bethesda. I asked them what are the advantages and disadvantages of being at a research institute. They said the first benefit was that they teach very little and are free for their research. They also said that government determines the direction of their research only up to a certain degree. These men had been professors in a great American university before going to N.I.H. They all felt that their freedom for research was greater at N.I.H. than it had been in the university. Thus I should like to ask whether or not we are facing a time in which high-level research has to be concentrated in some type of special non-university institute.

MORISON

To touch on another aspect of Mr. Cappelletti's remarks, the technical schools in this country were started by and large at a time when they answered a direct need—teaching people how to run machines. They have had an interesting history since. Five or six have remained at about that level, progressively becoming third-, fourth-, and fifth-rate as intellectual institutions. Two of them have become, as it became increasingly obvious that technology rested on science, more and more speculative within the fields of science and engineering. They both also have struggled hard to retain themselves as specialized institutions, I think mistakenly. Both would do far better to become general universities for two reasons: Their research would profit by the spirit of the university, and it is excellent for engineers and scientists today to be part of a general civilizing experience. If it were my decision, I would not even consider C.A.T.'s [Colleges of Applied Technology] now. Rather, I would consider altering existing engineering schools so that in time they would become universities.

In reference to another point Mr. Cappelletti made on the creation of separate research institutes, I think we could demonstrate that research does not flourish in this country as a separate enterprise; it has to be tied in some way to the functions of learning and instruction. Moreover, the dangers of a pure research institution becoming hooked to purposive research are very great. You might wind up as a government lab with diminishing returns in your intellectual life because you would increasingly be restricting your investigation to what was needed at the time.

ONUSHKIN

I should like to make some comments on the problem of utilizing market mechanisms in education. I am not against using some elements of the market mechanism (for example, economic stimulation) inside the system of planning, but I do not quite understand who, from Professor Ben-David's point of view, would be the customer—the students or the society? If it is the students, I am against such a market mechanism, because it will not create a situation of social justice. The students who are going to buy knowledge, or the services of higher education, in any case should know that they can use the knowledge that they are going to get; inevitably this means that they should have more or less clear perspectives. If they do not know whether or not they will be able to apply their knowledge, there will be a waste of scarce resources. Perhaps the optimal situation would be the combination of central planning and the initiative of different universities, but the customer should be the society as a whole—not just individuals.

BEN-DAVID

Another point on which I should like very much for you to say something is the relationship between the Academy in the socialist countries of East Europe and the university.

ONUSHKIN

In developed countries, there is a tendency toward universalization of higher educational institutions. These countries, and to some extent developing countries also, are getting more and more full universities. As Professor Parsons remarked, the problem of the combination of teaching and research is extremely important, and this combination of teaching and research is one of the main features of the modern university. I cannot imagine the university without research. Yet, at the same time, in some countries, like the Soviet Union or the other socialist countries, there are special organizations that are preferred in research work and are doing only research work.

Basic research and some applied research are being done in the Soviet Union at the institutes of the Academy of Sciences and at universities. The division of labor is not strict between the institutes and the universities because the people who are doing the research at the Academy of Sciences often also teach at the university. Similarly, many university professors do research at the Academy of Sciences. At the same time, there is a system of research institutions for applied research and development; these institutions are under different industrial ministries. The Academy of Sciences is responsible for the coordination of basic research; it works together with the Ministry of Higher Education.

The money for research at the universities comes from two sources: from the Ministry of Higher Education and from industrial contracts. The State Committee on Science and Technology coordinates the applied research and development for the whole country.

Until two years ago, all the universities and the institutions of higher learning were under republican Ministries of Higher Education. At the same time, however, we had the Union Ministry of Higher Education which coordinated the work and was responsible for the development of the whole system of higher and specialized secondary education, the high standards of training of specialists, methodological problems of teaching and research, publication of periodicals, textbooks, and so on. Two years ago, the Union Ministry of Higher Education decided to have its own universities and institutions of higher learning as experimental bases. The Ministry has chosen for this purpose a number of universities and other institutions of higher learning from among the best of their kind. These are Union universities, and they are the focus of experiments—curricula changes and so forth. The Ministry generalizes the experiences and invites the republican ministers to employ them. Universities under the republican ministries certainly can also go to the Union Ministry for advice.

KILLY

Medieval conditions still prevail somewhat in Germany. All our universities are financed by *Länder*. These states are coming to realize that a modern university is an expensive affair. Some are able to finance their universities; others are not. We are moving out of this situation because the legislature is trying to help the, so to say, underdeveloped *Länder* finance their universities by supplying up to 50 per cent of the financing out of federal funds. Although the university in West Germany is a state affair, there are quite a lot of other sources of income, both official and private. Among the official sources is the so-called Forschungsgemeinschaft, a current foundation of the German university which gets its funds from the Bund and is a self-governing body. This organization tries not only to finance research, but to emphasize certain fields and to even out differences that might arise. Foundations are not so important in Germany as they are in the United States, but the Volkswagen Foundation and, to some extent, the Thiessen Foundation are doing a lot of good things. All this is, as you know, not sufficient and we are trying to blueprint the future by means of two bodies: One body is the so-called Wissenschaftsrat, consisting not only of academics, but of public figures as well. This body has an efficient staff and tries to analyze the future but with, I am afraid, not much success. One of the main problems, especially in the sciences, is that up to 70 or 80 per cent of the financing comes from private sources, very often industry. These big industries cannot be controlled by the university and might influence the liberty of research in a detrimental way. We are trying to get this under control, but it is going to be hard.

One drawback should be mentioned. We still have an annual budgetary system, and this is absolutely nonsensical because, as you know, we need much bigger planning, much more time for planning. I think we can, however, overcome this difficulty within the very near future. I am personally involved now in an attempt to found a private university in Germany that will have a built-in buffer between its financial sources and its governing body. I am not sure yet whether we will succeed with it.

PIZZORNO

I should like to describe just one new way of allocating money to universities in Italy, for the general pattern is not far from the German one. The new way is the pattern of allocating money through the National Research Council. There are two ways, it seems to me, in which the method of funding is particularly important. First, purpose can change because of the structure that has received the money; second, the relationship between teaching and research can change.

The National Research Council is divided into seven or eight sections with a committee for each section. The allocation is decided by the general committee of the Research Council. Only full professors can apply for research grants; if you are not a full professor, you have to have the signature of a full professor to apply.

These funds are meant for research, but everyone knows a goodly part of these research funds goes to the universities. You receive the money for a research project, but a large percentage of the project's funds goes for general expenditure which is meant for the university itself.

SHILS

What is the proportion roughly?

PIZZORNO

It ranges greatly. In sociology, we are rather honest in our accounting.

Even though there is a central planning authority here, the allocation still flows through the channels of a baronial structure. Thus, money that should go to research directly is allocated by an administration that has nothing to do with the university as such. The administration that allocates funds for research is, in fact, weaker than the universities and, through the power of the baronial structure, some money which should go to research flows into the old university system. When you have such a strong local structure at the bottom, it does not help to have a central planning authority. You cannot go against this structure; after all, this local structure elects the committee and you have to keep up good relationships with them.

To speak to Mr. Shils' point about the multiplicity of sources of funds, the money comes from many sources. Although about 95 per cent comes from the state, there is always support from local authorities. I do not see that the university in Italy is much affected by the origin of the money.

SHILS

When I raised the question about the multiplicity of sources, I had two aspects in mind: on the one side, the protection of the autonomy of the university from excessive pressure from one single institution that would have its way with the university and, on the other, a tendency toward disaggregation of the university, with staff members using the university as a bookkeeping convenience. Of course, where the university is already as disaggregated as the Italian universities are, there is no *esprit de corps* anyway. Yet *esprit de corps* is necessary to a university to keep it focused on its central tasks.

GRAUBARD

Perhaps M. Bourricaud might say a few words on the situation in France.

BOURRICAUD

This topic is rather complex due to the present university conditions. I should like to start with certain little disputed points of fact. First, the *grandes écoles* are the core of the system. Second, at least until 1950, about 40 per cent of the students were registered in the faculties of law and medicine. Third, law was taught at the undergraduate level. Fourth, the research organization was more of the Soviet type than of the Anglo-Saxon type; research and university activities were clearly divided.

The university itself was run and managed according to the most tightly centralized pattern of administration. It is extremely difficult to understand how innovation or change could enter into such a picture, especially because the show was run not so much by the central authority, as by the corporate body of the professors—the academic profession.

Most of the personnel was financed through the state budget. The only way to get something new was to create new positions, although even then the procedure was rather long and complex. The initiative had to be taken by the local faculty or university group asking for the opening. The request had to be brought before the central agency, which would have to make that request in line with the budget and ask the local faculty to designate such and such a person for the post. The final decision would be made by the Ministry after a special corporate committee had been asked for advice. Thus, the whole thing was in the hands of the central administration on the financial issue and in the hands of the corporate group as far as personnel problems were concerned.

As far as research is concerned, basically two distinct branches of the state were interested. The central authority in the Ministry of Education provided the university with funds for minor research. The Centre Nationale Recherches Scientifiques was operated apart, though most of the people who were the barons within the university system were also the barons in the C.N.R.S. In the Italian system, research money has to be spent within the university structure; in the French system, it has to be spent outside. In sharp contrast to the Italian situation, no funds come from regional authorities. We have some territorial political authorities, but until now they have not made appropriations to universities. And finally, foundations are, for practical purposes, ineffective in France as the money they can spend is very small.

GRAUBARD

Mr. Onushkin, would you care to give us an idea of the situation in the Soviet Union?

ONUSHKIN

The number of students in higher education in the Soviet Union is 4,311,000; this number includes full-time students, day-time students, and evening students. There are three kinds of institutions of higher education in the Soviet Union. When I speak of institutions of higher learning, I do not include the specialized secondary schools.

The institutions I will describe are university-level institutions. First of all, there are full universities; then polytechnical institutes or universities; and third, specialized institutions of higher learning. There are forty-two universities in the Soviet Union. All the republics have their own universities, and some have several. There are 433,000 students at the universities, which means that about 10 per cent of all the students are in the universities. Despite this rather modest representation, the universities play the leading role in the higher educational system because they combine teaching and intensive research.

The universities are the only educational institutions that combine natural sciences, social sciences, and the humanities, and this is a particular feature of the university. In specialized technological institutions, for example, the social sciences and humanities are taught, but the research in these fields is not at such a high level.

The universities usually have faculties in the natural sciences, the social sciences. One of the important new features of the universities in the Soviet Union is that the leading universities—such as the University of Moscow, the University of Leningrad, and the University of Kiev—now have special institutes for improving the qualifications of the young professors or teachers from the other higher educational institutions. As you know, the young professors can carry the greater part of the teaching load of the universities. That is almost the universal situation and the reason why they do not have enough time to do research work. Thus, it was decided in 1966 by the government that it would be useful for the younger professors and, as a result, for the whole system of higher education to create special institutes to improve the qualifications of the young professors. As a rule, they spend half a year in such institutes. They do not teach at this time; they participate in discussions, do research, consult leading professors, and so on.

The polytechnical universities, of course, combine different fields of the natural sciences and mathematics. There are fifty-three polytechnical institutions in the Soviet Union.

The third part of the structure of the higher educational system is comprised of specialized higher educational institutions —technological, medical, juridical, economic, and so on. As a rule, these institutions concentrate in one definite field. Of course, the development of science and technology creates new fields; and specialized institutions, such as technological or engineering institutions, become more like engineering universities. The same is true, I should say, for medical institutions, which become more and more medical universities, teaching and doing research in different fields of medicine and in some aspects of biology, physics, and mathematics. The results of all these sciences are used in medical research.

There is a Ministry of Higher and Specialized Secondary Education in the Soviet Union, and this Ministry is responsible for the planning of higher educational developments in the Soviet Union. The Ministry is also responsible for the methodological part of higher education—the preparation of textbooks and so on. This creates a general curriculum for higher education which tends to be the same in all higher educational institutions in the Soviet Union. I should like to stress "it tends to be," for the level of training is not absolutely the same because it depends mostly on the quality of teaching staff, and this is not the same in different institutions.

Some of the universities, most of the polytechnical institutions, and most of the specialized higher educational institutions are under Union Ministry of Higher Education or under the corresponding ministries in the Republics. Some establishments of higher education are under a special ministry, such as the Ministry of Health or the Ministry of Agriculture. Some industrial ministries have their own institutions. Nevertheless, the tendency is for higher education to fall under the jurisdiction of the Ministry of

Higher Education. This means that some of the specialized higher educational institutions are under two ministries.

From this description, you can see the system of finance. Almost all the higher educational institutions are state institutions, financed by the central government or the republican governments.

The general rule is that all faculty members divide their time equally between research and teaching. Certainly this distribution of time is relative. Because universities are financed from the budget of the Ministry of Higher Education, university research is financed directly by that Ministry. But the universities also have the right to have research contracts with industrial enterprises. For some fields of research, these contracts are becoming a more and more important means of financing. Certainly, from the general point of view, both sources of funds are state sources because industrial enterprises are state organizations. There is a difference, however. The universities can choose the research projects for themselves, and they can get money for this research from an industrial organization.

One area to which a great deal of attention is devoted in the Soviet system of education is that of the continual improvement of the quality of education. The rate of scientific and technological development is so high that the universities face great difficulty in training people for the future and not for yesterday. Equally to be included in this domain is the same approach to the qualifications of the teaching staff themselves. There is also the question of the contacts between the higher educational institutions, research institutions, and industrial enterprises. The exchange of ideas in this area was one of the main foci of the reform initiated several years ago and still under way.

On one of the evenings, Dr. Pierre Grappin, the former Rector of the University of Nanterre, was invited to join the group. His account has such interest that we have given it in full, as he has revised it in fall 1969.

GRAPPIN

Nanterre was created in 1963 essentially because of the great numbers of French students. There were, of course, new principles, but the essential idea was to double the existing faculty. Those who made the new university did not make it exactly on the model of the old. The new principles that we have tried to apply consisted first in the materials that are taught. This is a very important point, as it could have a great influence on our future. In the course material, we gave a much greater place to psychology and sociology than has been done in any other French university. We had incontestably the most active departments in these fields in France. This policy responded to a real need, one for which there was a great demand among the students and one which corresponded to the character of the present world. We put the accent on the "human sciences" and the "social sciences." Furthermore, we tried to have a method of teaching that was simple and direct, close to the subject, and without constraint. In the first year, we were able to do this because we had enough professors in relation to the number of students.

It must be underlined that the environmental conditions for Nanterre were very bad. The university was a wholly artificial creation in a social context that was difficult to define. It was located in a wasteland; there were railroad lines, but practically nothing else except a few colonies of squatters living in board huts. In the first year, we were able to overcome these negative factors by giving to the professors and the students the feeling that they were in the process of doing something totally new, that they were building for the future. For this reason, we were able to obtain very appreciable results and without real opposition from the students. Thus, we were able to create a university experience which was the first of its kind in France and one which was often of interest to people elsewhere. We were able to unite the learning process and daily life for the students. We were able to create a place for the students to live near the place where the teaching went on. And we did this for a very large number of students—1,400 in 1965, half of them girls and half of them boys. In France this was a wholly new experience. The young people who were brought to live in this place had few conveniences. This was the milieu in which the new radical opposition was born. This opposition was born among the students who lived in this place because they felt themselves without guidelines; they felt themselves to be rejected by the society and that they were living in a sort of "slum" and not a normally constituted society. This represented the essential milieu from which the disorders sprung.

For me it was a tangible reality. The first violent incident that I had anything to do with took place in March, 1967, at the Cité Universitaire. On this occasion I saw for the first time someone I did not know. I did not know him, but an employee of the house next to mine knew him. He was to become famous in the years to come; it was Cohn-Bendit. He was there in order to direct the revolt against the regulations governing the Cité. They wished the abolition of all regulations, and because of the protests, the regulations have been completely abolished. At the students' residences at Nanterre, the students can do exactly what they wish. But in March, 1967, at least in my opinion, that element was completely new.

Actually, we had attempted more than one innovation at Nanterre. We tried, for example, to introduce an element that was completely new to the French university—the students living and studying in the same place. For this change, we evoked examples from the Anglo-Saxon countries, all of which you know well. We also wanted to involve student and faculty participation in this enterprise. This succeeded for two years, but was suddenly destroyed by the boys who lived at the Cité without professors. Only students live there and no professors. These students constituted a group that was no longer amenable to anything. Quite naturally they wished to make something from the situation they were in. Starting in November, 1967, a communal society was constituted there, as if on an island.

The elements of the second milieu in which the confrontation was organized, with a revolutionary affirmation, was the milieu of department of sociology and also of psychology. I saw the first signs of the revolt when torn-up lists of students were brought to my office in order to show me that from this point on it would be impossible to have any form of regulation or central authority over the teaching that would be given.

Exactly a year ago a large movement of *revendication* began among the students. It was concerned with professional questions, those issues having to do with the university. The students also complained that there were too many of them. In 1967 there were five thousand new students, and there were close to eight thousand students already enrolled. In 1964 there were only two thousand; in 1965 there were five thousand; but at the beginning of the term in 1967 there were close to twelve thousand. This is a large number of students, given the fact that many things were lacking: We did not have enough professors or assistants to teach them, and so forth. The movement, therefore, was concerned with professional questions, those having to do with the university. Its leaders arose spontaneously—at least that was my impression—and did the best they could. This movement lasted about eight days and then it dissolved; I am not quite sure why. There were very large assemblies where I spoke; there was a delegation of students to the faculty council, and we found, therefore, a way of making contact. We instituted a mixed commission, with both students and faculty, to consider the disputed questions. It was necessary to wait several months before certain other questions once more came to the fore.

When the revolt began again in March, 1968, what was most striking was that the leaders were no longer the same as those of the preceding months. Those who had led before—and whom I knew—had almost completely disappeared. A change involving the movement's very nature had taken place. In November, 1967, we had a movement within the student body that centered on questions concerning the university. In March, 1968, we saw an important, politically well-organized minority movement formed for a political struggle. Thus, the leaders were no longer the same.

In January there was a terrible battle. But in January, February, and the beginning of March, the revolt consisted of a relatively small number of organized students. By the middle of March and the beginning of April, when the students held public demonstrations or made speeches in the corridors or the classrooms, they had a very large audience of comrades. Then the movement became altogether different; a great number of students listened to speeches about razing to the ground the very principles of university education. From day to day, the demonstrations became larger, to the point where I think it would be inexact to call this a minority movement. At Nanterre a large number of students supported it, although probably for many diverse reasons.

The movement had several "natures" in the stages of its evolution. What was most striking about it was the facility with which a large number of young people—of diverse origins and of diverse aspirations—were veritably led to participate in it. They were fascinated by the idea of a social revolution and threw themselves into the movement without any kind of interior resistance, without often much sincerity, but with perhaps a kind of enthusiasm. This movement incontestably was originated by a very small number of people, but it echoed the thoughts of a much greater number.

CROZIER

I should like to add some comments on the change of mood M. Grappin has described. UNEF, the national union of students, had disintegrated, and there was general apathy. But small spontaneous groups were active on an informal basis, and the first leaders of the movement came from these groups. These leaders sought reform, but this was difficult to achieve as it would have disrupted the traditional *laissez faire* which meant a certain freedom for students. These leaders were the solid citizens of that community, but their efforts could not meet with any success because they were running against all the state machinery, because the university itself had no power of decision. Thus no solution could be found for them. The faculty and administration understood the situation, but our proposals were turned down by the Ministry.

Let us take the case of Cohn-Bendit. He could act as the victim himself. According to the compromise that was made, he was supposed to come into the third year. But since he had failed an exam the year before, he was put back into the second year. He became the representative victim and that was his first starting point. He was a revolutionary; he was agitating; he found an issue and won the support of the people who had grievances. He defeated the old leaders completely to the point that they just disintegrated and they became irresponsible. We tried in the sociology department to meet with them to discuss the situation, but they would not come to the appointments or they would disappear. We tried to meet them, but we could not find them; they did not represent anything that was real; and they were under attack from the extremists all the time because they promised a lot of things and nothing happened. After about two months of hesitation, the extremists picked up, and then the movement became something else entirely.

GRAPPIN

Last November the reasons for protesting were of a "technical" nature. The application of a new system of study made difficulties during the time of transition. These difficulties touched very few people. But the grievances they really had were much more vast, much more profound, and even these first grievances revealed the deep uncertainty felt by a great number of students about the French university system. For at least ten years, this system had remained quite extravagant; an absolutely uncontrolled number of people was allowed to enter, without anyone taking the trouble to know what they wished or what could be done with them. It is, in my opinion, a university system too far from its students.

One felt, at that moment, that these young people complained less for the precise reasons they gave than for a far more general reason—an all-pervading uncertainty. These events must be viewed with this understanding. Since January, 1969, the student movement has been led with much more political aims, in a tactic of constant provocation. They attempted to force the authorities either to abdicate or to retaliate. This movement was led, in a systematic fashion, by men who certainly had political experience and who were something entirely different than the young people who followed them. It is difficult to imagine that a young Frenchman, aged eighteen and a graduate of the *lycée*, could devise a political tactic so clever and so effective.

I was put in the position of having two choices open to me and only two: I could either annul my own position, my institution, my authority, or I could use force to enforce my power of decision. I did not wish to do either of these things, but I had only these two choices. The people who thought up this tactic were very intelligent. I do not think that the ordinary eighteen-year-old

French student is capable of concocting this. I assure you that the students I have known in the last ten years could not have conceived of a tactic like this because they were not interested in destroying the universities. Their revolutionary purposes did not concern the universities; quite a new kind of revolutionary tactic appeared among French students during the first months of 1969. These leaders felt that the French university system belonged in a museum. I understand perfectly well that the authority that reposed in me, as the head of a liberal university, depended entirely upon mutual consent, but I had no recourse, except force. From the moment people who respect absolutely nothing of a system's fundamental principles install themselves in that system, they can do anything they wish. They have absolutely no obligation to it. They pay very little—eighty francs per student. The university has no recourse: It can neither in practice limit the number of students nor forbid them to enter or to present themselves for an examination. This system depends entirely on the good faith of those concerned with it. From the moment that you find yourself confronted with people who—for whatever reason—have decided to ignore these rules, authority has no grounds on which to stand.

PIZZORNO

You said that in April the character of the *revendication* was completely different than it was later when it assumed a political aspect. In what sense do you mean "political"?

GRAPPIN

The students became radical in the etymological sense of the word. It was no longer a question of university regulations. They were attacking the fundamental principles on which the university—the society itself—was founded.

PIZZORNO

How did this attack manifest itself?

GRAPPIN

I talked with the students a great deal, even during the times of the terrible battles. They said to me that the university—all society —had to be made over. They offered as proof the following argument: You, who are an esteemed authority in the university, employ the methods of repression against us. These, I told them very sincerely, were the repressive measures they had asked for.

But the result was the radicalization of the movement. University society is the most open and the freest of all the forms of society I know. It is the form in which authority is farthest away and the most liberal; but the students demonstrated to themselves that even this society was founded on repression. In this sense, one could no longer speak of a possible resolution to the problem; the fundamental principles of society were involved. On this subject there are many possible variants: The war in Vietnam plays a role, as does the imperialism in which we are engaged. The crux of the idea is that we are a society that will use force and repression for the purpose of limiting each individual's particular talents. From that point of departure, they went on, for hours and hours, about individual creativity, a new method of teaching that would reinforce this creativity and not limit it, a new society that would be a place for creative people.

You have certainly heard discussions of this kind. The political element comes in. There was a small political faction in the large assemblies of students that was not in accord—Marxists, Maoists, Trotskyites, anarchists. The political transformation of this movement came when the movement itself became a radical critique of the society. This was the dominating theme in the end and was, incontestably, anarchistic.

ZUCKERMAN

In a way, M. Grappin answered the question I was about to put as he discussed the transformation of the attitude of the students and the emergence of a radical political objective. Given the existence of student upheavals in many countries, would you say that there is some sort of diffusion of this defiant kind of unrest? The London School of Economics is not yet out of its troubles. I should be very surprised if the political motives inspiring the young people in England were the same as those disturbing the young people in France or in the United States.

GRAPPIN

The elements most active in organizing the revolt in France were the politicized elements—the students who belonged to political groups.

KILLY

May I try to explain the difficulty? I am afraid that to an Anglo-Saxon mind this revolt is not worth being called political because it is wholly unreal. It is not of this earth.

GRAPPIN

It is.

KILLY

In your country and in mine, a strange or dangerous contamination has taken place between the political and the utopian, between the ideological grievances and the real ones.

BRIGGS

I first came across the term "imperialist-capitalist complex" in February, 1968, around the same time that the university grievances began to take political form in France. In England the language of the upheaval was exactly the same, as were the forms of organization, the talk about organization, the references to committees, to general assemblies, to coordinating committees. This was so even though only a very small minority belongs to the rebel group.

ZUCKERMAN

There is no other language?

BRIGGS

They have a language quite different from that of the 1930's. I should like to know just when communication at Nanterre between the university authorities and the students began to be increasingly difficult. I take it that in November, 1967, when the issues were essentially about the structure of the university and student grievances, there was communication; but that when the political issues began to become more prominent and different factions began to develop, each with slightly different approaches, communication began to be extremely difficult.

GRAPPIN

In November, 1967, communication did exist. M. Crozier underlined this. We not only had communication, but were able to arrive at an agreement rather easily. We had such a rigid university system that, so far as our university was concerned, we were not able to take the measures that we had been forewarned that we should take, measures that the rest of the French university system did not consider necessary. Communication was cut because of two facts: On March 5, 1968, I was besieged in my office by about fifty people, who had come to bring me a paper that I was supposed to sign, a paper that provided guarantees for freedom of political expression without reprisals. Furthermore, I was to give guarantees to the effect that there would be no reprisals in the grading of examinations.

I said I would never sign anything under duress, and that they should go back where they came from. I informed them that if they sent me an elected delegation, with men who could tell me whom they represented, I would be perfectly willing to discuss the issue. One of them replied: "From now on we will use guerrilla tactics against you. We will never tell you who we are; we will never be electorally responsible; we will have only secret and anonymous counsels." From that day on, the situation was impossible to control.

ZUCKERMAN

I should like to ask where the professors were during the month of March?

GRAPPIN

I will tell you something very funny. On March 5, two professors were nearby—one a professor of geography, the other a professor of French. The professor of geography has always testified that things happened as I have told you; the professor of French has forgotten everything.

MORISON

Much of what I have to say has, I think, already been touched on. We have spoken about the impact of research and also spent a lot of time on students, debating how the variety of students might cause the universities to alter their structure. We have also discussed other pressures of equally persistent and powerful kinds that may cause universities to rethink their structure and mission.

I should, therefore, like to begin my remarks by responding to Mr. Cappelletti's question of yesterday about the wisdom of having separate research institutions. Professor Parsons knows more about this than I and has more data, which I hope he will give. I can report, however, that historically the best work in research has been done within universities; for as long as the investigation of ideas was taken to be primarily an end in itself, it was very useful to have students—not only graduate students, but undergraduate students. The exercise of converting research findings into ideas that could be transmitted to undergraduates gave a kind of validity to the work. If the training of graduate students is largely intellectual, the education of undergraduates has a moral component. The university is endeavoring to educate knowledgeable and therefore more useful citizens. The existence of this moral obligation—that is, to turn out more effective citizens—and the awareness of it anchored and stabilized research. For this reason, the history of detached institutions without students has been bad, on the whole, in terms of output.

I have some reservations now that this situation need necessarily continue in certain areas. My data are drawn from a special case, the Massachusetts Institute of Technology, where I was for twenty years. Increasingly while there I wondered whether the undergraduates were getting a reasonable return for their investment; I am pretty clear that they were not. The Institute, I should say, was without any qualification the most exciting and vivid intellectual community that I have ever been in. Most of its work was done in connection with the immediate needs in the society. It is infinitely responsive to what is going on, and the validity that the older universities found in converting their information for student use, M.I.T. found in converting information and research into immediate application in "the real world." It sustained itself effectively as a general investigating institution needing graduate students, but not so clearly needing undergraduates.

All universities have increasingly been under greater pressure to develop and refine ideas of immediate use. On the basis of my experience at M.I.T., the first question I would raise is whether or not you can sustain yourself in good health over a long period of time doing what I would call *ad hoc* research. The vividness and the excitement of that life I cannot understate, but it lacks a certain philosophical dimension, which I think is not simply confined to the Massachusetts Institute of Technology. One has to think rather more than we have of the meaning of the kind of intellectual work done in the most important fields today in the classical universities. What changes may be made in the quality of their life, in the character of the men that are doing the work, and in the general conditions in which knowledge grows?

There is a second problem not immediately so pressing or so interesting philosophically, but one that must be met in time. The projected budget for the year 1976 in high-energy physics in this country is $500 million. If you take this figure and then add to it some of the current skepticism about the returns from such research, you begin to raise questions of priority. Where, for example, does biology fit in relation to high-energy physics, in relation to geophysics, in relation to psychology? These are questions we have not had to ask ourselves up to now. Who will make this determination of priorities? Where will such determinations be made and on what criteria? We have no adequate data to enable us to make good decisions about putting funds into another great accelerator or into microbiology.

Finally, as has been reiterated over and over again, the money that is spent on research comes increasingly from federal sources and to a great extent for specific areas. What distortions will be created in the nature of our general intellectual life by this focus? As a member of the humanities, I believe this question raises issues that come close, ultimately perhaps, to the survival of humanities: The significance of history and of literature is harder and harder to discern in the present situation where money is available easily for certain kinds of things and not for others. I cannot leave this topic without saying that to the extent that the humanities have borrowed the procedures of the natural sciences and research, they have done themselves a profound disservice. One reason that the humanities account for as little as they do today in the university is the character, if not the quality, of the research and the meanings of their findings. Universities must think to an extent that they have not yet about the nature of research in those subjects that supply a general or cultural context. How far can these fields appropriately borrow from the natural and physical sciences? What kind of graduate training is appropriate in the humanities? I think we are sending out poor practitioners to deal with the questions of value in an industrial society by virtue of the kinds of things we are doing in the humanities.

Particularly in America, which has always been interested in application and engineering rather than basic science, we may be using up the reservoir of fundamental ideas. If not that, we may at least be in danger of training people who are not interested in enlarging that reservoir; we may simply be producing a group of practitioners and appliers. We may run the risk of working ourselves out of any inherited cultural context.

MEYERSON

I should like to try to tie Elting Morison's points to Ed Shils' comments yesterday on the disaggregation of universities. It is constantly claimed that the kinds of research funds that go into universities remove the academic man from the student, from the classroom, and from university loyalties. It is argued that research funding creates a set of loyalties that shift the academic man's fidelity to his research guild or to the granting agency. In the United States, relatively few universities are doing the expensive research. Probably a score of universities has the bulk of the doctoral students and the bulk of the federal research funds. Thus, the disaggregation, if it is taking place, is surely not taking place because these research funds are going into the universities. We see this disaggregation in endless institutions and not just in those institutions that receive the bulk of the research funds. Why do faculty members see their primary allegiance not to the university but to their guild? If this description is an accurate one, how long has it been accurate? When did this change take place?

We need case studies in considerable detail to decide what the transformation has been. Indeed if there is evidence of such a transformation, it is probably the result of the rise of graduate education rather than of the rise of research. If each of the 500,000 college and university teachers in the United States wrote a book every fifth year, more books would be published than are published in the English tongue. Obviously this situation has not occurred, and thus it is not writing and research activities that have caused the disaggregation. Nor am I not prepared to say that the disaggregation is all bad.

Finally, to touch on the difficulty of making large-scale allocation decisions: For several years some of us have been involved in helping to get started the two-hundred-billion electron volt accelerator in the United States. This will be the world's largest accelerator and will dominate high-energy physics. But why high-energy physics? Why this huge allocation in that direction? It is essentially because of the tremendous reservoir of goodwill that the high-energy physicists have built up for themselves. With the long history of nuclear accomplishment, how can the physicists possibly be wrong? Thus, we are breaking ground on this accelerator, and no one can predict the order of new knowledge that it will provide for high-energy physical research. Some high-energy physicists themselves feel that they have probably gone about as far as they reasonably can in uncovering new knowledge. We see the problem of a marginal determination here that quickly becomes a political determination, a political determination made not by the so-called industrial-military complex, but rather largely by the guild of high-energy physicists themselves.

ZUCKERMAN

I should like to address myself to the primary question which Elting Morison raised concerning the virtue or lack of virtue in separating research from university education. In general, on the basis of British experience, I would agree absolutely with everything that he has said. At this moment in time, research in Britain is carried out in universities and also in a number of institutes, paid for by government, which are devoted to research and not at all to teaching undergraduates. Some of these institutes are also concerned with development. There is some pressure at the present moment to attach some of them to universities or to give them certain university connections.

Let me give you an idea of the scale of the problem in the United Kingdom. Our scientific civil service employs some 12,000 professionally qualified scientists and engineers. Assuming a staff-student ratio of ten to one, this is equivalent to a student population of 120,000—something like half of the university population of England at the present moment. In other words, I am talking about a body of scientists as big as the body of scientist-teachers in the universities, if not bigger. They vary greatly in their scientific quality. The institutes lack the free flow of new young people that one experiences in a university department, and therefore a major stimulus in research. A colleague of mine, long since retired as head of a major department in Cambridge, used to judge the relative merits of candidates for academic honors by asking how fast the results of their research work were becoming incorporated in the curriculum for undergraduates.

Some of our governmental research institutes are extremely good—the biological ones, in particular, are good. These are small, and they all have to maintain close links with universities. The National Institute of Medical Research, for example, has a very close link, and in terms of achievement does much more than the others. These others, however, house the bulk of the army of twelve thousand professional scientists and engineers—institutions such as our National Physics Laboratory, our aeronautical research establishment, our radar establishment, and so forth. These institutions are not just devoted to applied work. Something like 20 per cent of their budget goes to basic research of one kind or other. Yet in general the scientific achievements of these people are poor—measured in terms of recognition—compared with the corresponding achievements of people working in universities or in institutions with close university links. I think secrecy and isolation are responsible here. Harwell, for example, thrived while it was extending and developing basic propositions of nuclear physics which had been developed mainly in universities, but when it went into the business of technological development and turned from the kind of fundamental research which one associates with a university environment, its role became so blurred as to require a new definition of its mission.

PIZZORNO

At least two important points were made by Professor Morison. One is the relationship between research and teaching, and the other one is the problem of incentive. I should like to describe two kinds of experiences that I have had: one in an institute more or less of the kind Sir Solly mentioned and the other in a graduate school. For seven years I worked in an institute that was supposed to do only research in applied urban studies; it had about seventy or eighty degree-holders in sociology, economics, urban design, architecture, and statistics. This institute was supported by local authorities, and rarely could one see a group of intellects of this distinction working together in an institute in Milan. The experience was interesting and distressing at the same time. The main purpose of this institute was to do research on the bad situation of the Italian cities, which meant a practical, useful type of work. Most of the institute's members were university professors. We had the right to publish everything we produced. Our official public was the public authority, but we did not care at all for their judgment. We still had the university public, our guild. We were in a very uncomfortable situation, because our real prestige came from the guild, although the immediate judgment came from the political world and the bureaucratic world to which we gave our results. The difficulties might have been tempered by putting some function of teaching into the institute.

The second experience I had for just one year at the graduate school in Milan. About thirty students were collected from all over Italy; all had done some scientific work. They were so bright and so demanding that they became the public; they became an alternative to our guild public. This was the ideal situation for teaching. Some of us devoted 100 per cent of our time during that year to this new public without doing research work just for the pleasure of being recognized by this public. At least in the first half of the year, we were pleased with this alternative public, but the public itself was not so pleased. The students wanted to have some sort of outlet of their own. They became somehow more at ease when they began to see the possibility of conducting research of their own—and not just for the purposes of acquiring information.

SHILS

I agree very much with what Professor Morison and Professor Pizzorno have said, but I think one aspect has been omitted. I hope no one will misunderstand me when I speak of the erotic aspect of the teacher's relations with students. The presence of young people—full of life and intellectual vitality—is extraordinarily stimulating quite apart from the intellectual contact. When people praise some members of the academic profession for their vitality, they often do not appreciate that much of that vitality comes from living in the animated atmosphere created by lively young people.

PARSONS

There are a number of points that I should like to speak to. First, in relation to Mr. Morison's statement, I have been engaged in a study of academic professionals in this country, and I think we have some interesting attitudinal confirmation that the combination of research and teaching is highly valued. [Professor Parsons's findings have been published in Talcott Parsons and Gerald Platt, "Considerations on the American Academic Profession," *Minerva*, Vol. 6, No. 4 (Summer, 1968).] The complications of interpretation that the data pose involve the stratification of the American system. People in the less prestigious institutions, generally speaking, have little opportunity for research, and many have very heavy teaching loads. It is not at all surprising that these people want more opportunity for research than they have. As you go up the scale, however, one finds among many scholars on fully or primarily research appointments a desire for more opportunities to teach. The general trend, as Mr. Shils noted yesterday, is for the profession as a whole to want more opportunities to do research, but this is largely a function of the stratification and the historic anchorage of the system in the teaching function, rather than of the devaluation of teaching in general. This phenomenon extends not only to graduate teaching, but also to undergraduate teaching. There has been a good deal of talk among the intellectual public about the flight from undergraduate teaching, but our evidence does not bear this out. There is certainly not an exodus from graduate teaching. But across the board, the respondents wanted their administrative loads lightened. Ideally they wanted to cut in half the amount of time they devoted to administration; this was true in all categories of institutions.

There is also a strong tendency to incorporate the whole range of the intellectual disciplines in university faculties rather than to specialize in one section. M.I.T. has moved from a technological institute to a general university with a technological emphasis. It has first-rate departments in political science and other social sciences. The California Institute of Technology is now also moving into the social sciences. You do not get in the United States schools of humanities or even schools—like the original conception of the

London School of Economics—focused in the social sciences with no natural science and little in the humanities. You get a strong tendency to the total range.

You also get another tendency, which may be motivated partly by guild considerations, to pull the professional training schools closer to the universities. This phenomenon started with the classical professional schools of medicine and law, but is now increasingly true of those in business administration and education too. I remember a former dean of the Harvard Medical School telling a story about President Eliot of Harvard. Shortly after his accession to the presidency of the university, he had the temerity not only to attend a meeting of the Faculty of Medicine, but to insist on presiding at that meeting. There was utter consternation among the medical people. They used the name of Harvard University, but they considered the medical school a totally autonomous thing. The Eliot administration brought the medical school into the university, which is a common story of professional schools in this country.

It is my impression that the strongest universities have gained greatly by being able to build on strong undergraduate colleges, and they have preserved the strength of those undergraduate colleges by and large. The University of Chicago went rather far at one time in de-emphasizing its undergraduate college and then came back again.

SHILS

It was not a de-emphasis of undergraduate education so much as it was a change from general education to preparation for more specialized study.

PARSONS

We have, I think, a cluster of aggregations in the American system that is in many respects different from the continental European system and more like the British. These aggregations are very deep-rooted in the system, and perhaps counteract somewhat the disaggregation, of which research entrepreneurship is probably the most conspicuous case.

MEYERSON

I accept Talcott Parsons' research findings, but would like to suggest that he would have discovered something vastly different as little as five years ago. For example, at one university, there were ninety-some research centers and institutes with people frequently promised a professorial appointment that could lead to an attachment to a research center or institute with little responsibility to students, undergraduate or graduate.

PARSONS

In our new sample we have data on Berkeley that have not yet been pulled out.

MEYERSON

But there are other examples of similar breaks in local fidelity. The events that started with Berkeley in very large measure have caused a tremendous sense of guilt among professors in America. Thus, throughout the country we find a widespread shift toward undergraduate teaching. The pattern of avoiding undergraduate teaching is now changing at university after university.

PARSONS

It is, of course, a variegated pattern. I have one sociological colleague who has made the assertion that nobody has ever assumed the burdens of parenthood voluntarily; of course, this is fantastically untrue. But what is motivation? It has to do, on one dimension, with the internalization of the role of the other in a dyadic relationship. Every human being has been a child and the child of particular parents. He has, in turn, internalized the role of parent. When he reaches the age when it is, in fact, feasible to be a parent, there is a strong motivational base for actually assuming that role. Every university or college teacher has been a student, and although by no means all students become teachers, among those who do there is a built-in set of needs to assume the teacher role that they experienced in relation to their own teachers when they were students.

MORISON

I want to reinterpose myself to sharpen and restate a point that I hoped that I was making. I feel that pressures are being brought to bear on the university today making increasingly difficult the task the university used to do in fundamental research. I am not to be put off by Mr. Meyerson's observation that federal money goes mostly to ten or fifteen institutions. They are the ones that set the tone for the country's intellectual life as a whole. The people in those determining institutions are subjected to strains that no human being can easily withstand—particularly when they enable them to do very interesting work. But I see a real change in the character of the people who are determining our intellectual life. They are energetic, interesting; they are people concerned with problems they believe to be terribly important; they are very bright and skilled in operations. They are a new kind of man, nervous and operational and working to the moment. And they set a tone that infects not only their own areas, but the whole spirit of the university in a way that makes me nervous about the task of protecting the quality of intellectual life in areas other than the ones in which they work.

CROZIER

We do not have in continental Europe the attitudinal aggregation that Professor Parsons has described. We have separate institutions, and there is difficulty in getting any interchange. I have been in a situation where I have suffered from not having contact with students. I could have two roles, but these two roles could not overlap. At one time in the week I was a professor and during the rest of the week I was a researcher. It was absolutely impossible to mix the two roles. If people say this is possible in France, I would challenge them. Thus, not all university systems would be amenable to the mix of research and teaching that has been advocated here. The continental European universities certainly would not be conducive to it.

We should be interested in the way the system as a whole operates. There are contradictions in any kind of institution. What is important to understand is how an institution can operate in a fruitful way regardless of these contradictions. I agree with Mr. Morison on what should be the character of basic research; on the other hand, the new roles of the operational leaders are extremely important. Behind much of this applied research, which has a lot of inconveniences also, is the challenge of the environment and the challenge of young people.

One of the achievements of American universities has been to keep the general system working and cross-fertilized in such a way that new roles develop without losing the old kinds of achievement. They manage it by recognizing the contradictions in the allegiances within the various groupings. Modern life means a lot more complexity. We must have our cake and eat it too.

BRIGGS

I am doubtful myself whether the right word is *contradiction*; perhaps *tension* might be better. It seems to me that there always are tensions in the relationship between teaching and research. Sometimes those tensions can be creative; at other times they can be positively frustrating. I am relieved that Professor Parsons laid emphasis on the aggregating elements in modern universities, because the disaggregating ones do get talked about. On the whole, however, the aggregating ones have been quite obvious. For example, in Britain when the new universities were being created, people who might previously have thought of themselves as being associated primarily with the guild began to be drawn into new sets of relationships because of the special identity of the institution that they were helping to create. The particular process of bringing people together, including people from research establishments, has produced extremely constructive interchanges.

The quantitative expansion of university places in Britain has been associated with more emphasis upon the teaching role than there ever was in previous periods. This is a difficult subject to generalize about. It seems to me that the position changes almost every year. There are always tendencies and countertendencies. I have noticed one or two tendencies that may be operating in the opposite direction from those which have been brought out in Professor Parsons' research on America at the present time. There are people in British universities who are saying—and I have never heard this said quite so strongly before—that the presence of young people is not so much a stimulus as a positive disadvantage in carrying out their own lines of inquiry. This is said particularly in some of the social sciences. There is quite a strong feeling that economists previously teaching and doing research in the university would prefer to be associated with an institute created, for example, to study economic development. This tendency is also beginning to appear in the humanities. I have heard historians wish to be free of the rough and tumble of universities at the present time. I have heard comments that during the next ten years there will be new balances in Britain between research and teaching which will involve new institutional arrangements, and that these patterns will break quite sharply with the past. We have never had the disaggregation that has happened in the continental situation. The particularity of institutions has great importance in Britain. I have not seen this sense of institutional identity breaking down much.

Sir Solly talked about the sterility and the immobility of many people in the research establishments after a certain age. The important problem here is the mobility between these institutions and the universities; there must be some kind of free passage. If you have an expanding educational system, the free passage is relatively easy. If you are not expanding so fast, you begin to encounter difficulties. Many of the people who went into these institutions were probably less interesting at the point of entry than some of the people who went into the universities. It may well be that we have got to work out systems where people can be taken out of universities and put into research places for periods of time. We need places where people can recharge.

SHILS

If a mermaid were to leave the depths of the ocean and go to live in London, she would think it was extraordinarily dry. Asa Briggs lives in a university culture that is extraordinarily unique for its preoccupation with the training of undergraduates and its devotion to the training of undergraduates. Even if there appears to be a slight falling off in that, the level of concentration on teaching even among those who allege that they are not interested at all in teaching is extremely high in England.

BEN-DAVID

If one compares Professor Parsons' study with those done on the situation in England, it is clear that the university professor there considers himself to a large extent as a teacher. Quite a few professors in Britain hardly consider themselves as researchers. The whole composition and self-definition of the professorial role are different in England from what they are in the United States.

I should like now to return to a point Mr. Morison made about the superiority of the university as a place for research and add a few points to it. He spoke of the philosophical depth of the research. By having to organize a course, you have to put your own work into a much broader perspective, and you have to philosophize about it to some extent. You cannot organize a field of study for teaching at a high level without having some sort of philosophy about it. You may not be compelled to do this if you are only a researcher, even a brilliant one.

The other point I should like to make concerns the unity of the various disciplines within the university. It has often been stressed that the unity of knowledge has been falling apart for the last hundred and fifty years. It probably has; nevertheless, one of the great advantages of a university is that this unity has not entirely disappeared there. All creative people I know in any field have contacts across disciplines. It seems therefore that the combination of research and teaching as developed in the best universities has been, so far, the most fruitful way to conduct research.

There is, however, a kind of technological research which is creative and vital, but for which universities provide little incentive or opportunity. Many of the publicly financed research institutes established for the promotion of useful knowledge in fields with potential applications are not very successful in this respect either. As has been pointed out by Sir Solly, some of these institutions tend to become stale and unproductive. Perhaps they should never have been established to start with. But all this leaves open the question: What is the optimal setting for creative research in technology?

MORISON

There is a puzzling aspect here. In a sense, engineering development, which is more difficult than scientific research, combines both intellectual quality and the instincts of an artist to organize random elements into a single situation that will really work. This turns out to be almost impossible to teach except in the classic artistic apprenticeship way; it cannot be taught to large crowds of undergraduates.

KILLY

The deplorable state of the continental university has been mentioned over and over again. I would be willing to give up almost everything of Humboldt's teaching except this tension or interplay between *Forschung* and *Lehre*. The second experience Professor Pizzorno described was for him an extraordinary experience. To my mind, it should be the normal experience of high-quality university work. The malfunctioning of the continental universities can be traced to a great extent to the disruption of the very sensitive balance between teaching and research. The productive tensions such a balance might produce are destroyed from the beginning because continental universities are places of anonymity. There is no relation whatsoever between the so-called teacher and his so-called pupil, and there could not be a satisfactory relation for the obvious reason of numbers. As long as we do not succeed in getting rid of this climate of anonymity, millions of marks will be spent in vain.

I was impressed by the obvious changes in the Soviet system, which are changes to the good and intensify the tension between research and teaching. Remembering what Mr. Parsons said about the psychology of teaching, the main reason for this combination of teaching and research is for me not a psychological one, but a very material one. I cannot conceive of any productive work being done in a humanistic field without a colloquium, without the criticism of students. I could not live without my four or five assistants, and I greatly deplore that they are being alienated, not to say devoured, by their teaching loads.

The humanities simply cannot exist in an anonymous academic society. But there is another reason for the low ebb of the humanities which might be specifically modern. Only in rare instances do the humanities produce operational leaders.

PARSONS

I think that Mr. Killy has made a most interesting statement. I just wanted to say, first, that I do not think for a moment that the psychological factor I mentioned stands alone. I think Professor Ben-David made an important point in speaking of the pressure that teaching gives to deepen, broaden, and clarify thinking. I have had occasion to be concerned with sociological aspects of law, and I have been much impressed with the parallel between teaching and the procedure of adjudication. The courts must adjudicate, with very few limitations, any dispute that is brought before them; they cannot stay within a closed world of deductive propositions from a few premises, because they have to face new problems all the time. This is also one of the main functions of teaching; bright students will always ask the difficult questions, and you are not allowed to evade them. This forces the teacher to take into consideration things that he would otherwise have neglected. The motivational reasons why academic people are willing to expose themselves to the punishment of dealing with really bright students are another aspect of this whole question.

FULTON

I started as an undergraduate at sixteen in a Scottish university. People came there from very mixed backgrounds—from homes that had nothing but the Bible and from much more sophisticated families. You were suddenly confronted with the great master. He produced an astonishing proportion of scholars by bringing the light into their eyes at the right time. Then I went to Oxford at a time when the scholarship by modern standards was rather amateurish. The teaching was student-oriented, but there were still big men around. It is not so much that this teaching orientation has been lost, but that scholarship itself has become vastly more professional in approach. People began to realize that to get a chair in another university the only thing you could transfer was your writing. The systematic development of universities did, in fact, cause teaching to be regarded with a much more professional and guild attitude. People say they are interested in undergraduate teaching, but they are interested in undergraduate teaching as a reflection of postgraduate teaching.

I suspect that a great deal of the trouble in science has arisen because the people who are teaching are the people who respect the Royal Society imprimatur. That means they are addressing themselves to 10 per cent of their students; the other 90 per cent have got to undergo being taught as if they belonged to the 10 per cent and they know that they do not. These students leave the university feeling that they are failures. This situation cannot continue without creating terrible problems. I feel very much in sympathy with Mr. Killy, and the discussion about the continental universities reinforces the point that the teaching that goes on inside the university is not for the direct benefit of the students. The challenge for the university today is to pick the undergraduate course and revise it. It is terribly difficult to be a springboard for the postgraduate training of the next generation's scholars and at the same time to equip undergraduates so that they can make their contribution when they go out into the world. These two functions must be preserved; otherwise, in the so-called tension between research and teaching, the teaching is unconsciously giving way because it is becoming a reflection of postgraduate teaching. A great many university students are not going to be the scholars of the next generation. The greatest tradition of European universities has been that they have taken seriously their concern for these students.

DUNTON

As I understood Professor Morison, the pressures on the university are making it difficult for the university to do fundamental research, to speculate about big ideas. He put the blame mostly on the public demand for *ad hoc* research, as expressed largely by big federal grants for that kind of research. I suggest that a fair part of the blame may lie with the humanists and, to some extent, the social scientists themselves. They have been inclined to award their prestige for a paper published on a new bit of information about Chaucer. We certainly want to keep adding to that stock of knowledge, but surely the important thing is for humanists also to think about the big ideas, about values. There is the desire as well as the money for *ad hoc* research. But there is along with this an enormous growing demand from the public, particularly the younger generation, for the other thing—for speculation and talk and investigation of big ideas.

Enrollments in the sciences are tending to go up not nearly so fast as enrollments in the humanities and the social sciences. And I suggest that most young people studying in the social sciences may be disappointed if they only learn the methodologies of behavioral study. The humanities and the social sciences do offer an alternative to *ad hoc* research, and one that many students are looking for.

CROZIER

I agree very much with Mr. Dunton's plea that the humanities and social sciences address meaningful problems, but I fear we may have been misled in trying to reconstruct the past with the feeling that we can do so by reconstructing the parts. We should not be seduced into trying to retrieve old relationships and the old kind of content. We must reconstruct the past to innovate.

MORISON

When I went to the university, I was taught by a group of men in the humanities—literature and history primarily. They looked upon themselves, although well-trained scholars, as protectors of the old high culture. Imbedded in that high culture was a set of moral attitudes, and these professors were prepared to take positions—that Cromwell was a bad man or whatever. From this I learned a great deal about a set of values that then related to the world around me; I also learned some historical data. In recent times, this position with respect to the old high culture has been difficult for anybody in the humanities to assume for two reasons. First, the celebrated analytical method of science began to infect the humanistic fields, and humanists began to concentrate on refined analysis; they were trained this way as graduate students, which tends to kill some of the moral imperatives or the impulse toward moral imperatives. Second, the cultural arrangements that those people taught me no longer related so directly to the conditions of society. Thus, you were in danger of becoming either the purveyor of a past that did not so obviously relate, or you were being soft in not using the critical and analytical procedures that were making great headway in other fields.

MAEDA

In the system of higher education in Japan before World War II, we had relatively small classes in the humanities; now we have mass general education. There are four hundred students in philosophy classes, and you lose the intimacy that you had before. Even people outside the university have begun to realize that the student is not a real student under such conditions. He certainly has little interest in philosophy given this experience. So we are struggling to get back to the system of smaller classes.

MEYERSON

The dissident young have what they like to call the politics of rudeness, and I should like to borrow a leaf from their book for a while and be their spokesman. Since no one else obviously intends to do so, I should like to put on the table the kind of charge that they would make to us. They would say that we are asking them to do as we say rather than as we do. They would say the discussion we have had for the last couple of hours would do justice to the late Cardinal Newman or to Humboldt. They would claim this is all liberal garbage, although they would use a harsher and ruder term. They would ask where to find that blend of research and teaching we have been discussing. Who gets rewarded at the universities in all of our countries? What do we transfer to another university? Not our teaching, but our publications. Teaching is a capacity that is judged only by hearsay. Prestige rests on publication.

I have had delegations of Oxford students visit me and say how much they resent the tutorial system. Indeed, they resent the very erotic quality Professor Shils mentioned.

SHILS

They want that atmosphere regardless of what they say.

MEYERSON

Be that as it may, we will ignore this rhetoric of the angry young at our peril. The rhetoric proceeds in many fashions: Science and technological research presumably are geared to immoral ends. They claim that we engage in these immoral activities and reward only those who have publications rather than teaching skills. They go further and condemn our humanistic bias as the bias of an elitist group concerned with the past, concerned with kinds of humanism that may have been appropriate to the class structure of the Italian Renaissance, but hardly to the contemporary world. They claim it is the bias of the verbal against the sensual; a bias that ignores the visual, the auditory, the other senses. I am hardly saying that Tokyo National University is closed because of the discrepancy between what we say and what we do. Nevertheless, substantial numbers of students—and not just the most politicized—have a deep concern with an educational experience that they find very wanting.

SHILS

The students may desire a set of values rather than factual knowledge, but can we give this to them? Do we have it to give to them? Are they entitled to want it from us? The students demand all these things, but is it within the possibilities of earthly existence that they be given to them? It is certainly difficult to make young people composed and restrained in their expectations of life. We live in a most ridiculous age. The utopians from the sixteenth century to the nineteenth were in a few instances thoughtful, wise, benevolent men who wrote little essays; they had a few followers, not very many. But this contemporary utopianism has taken possession of a generation and that is very difficult to satisfy. Those who are most dissatisfied are in fact dissatisfied with our "values" so that expounding "values" to them would not please them at all.

They study social sciences in increasingly large numbers. I myself think—although my friend Professor Parsons probably disagrees with me on this—that it would be much more desirable if they did not study social sciences in such large numbers. What do they get when they study the social sciences? They come for bread, and they are given gas, which sends them into convulsions. Who are their heroes—Marcuse, the late C. Wright Mills, Eldridge Cleaver. They do not want the stuff Professors Parsons, Ben-David, Crozier, Bourricaud, and I give to them. They want something more exciting, and we are not in a position to give it to them.

We have talked about the burden put on universities by the large numbers, but we have not raised the question as to whether we can supply the teaching personnel. It is not just a question of what is called high-level manpower planning. If it were, we would only have to decide that we need so and so many teachers in this category, and so forth. Even if you could provide the magnitudes desired, can you produce the quality that is desired? I am not a romantic about the marvelous relationship between Mark Hopkins and the boy on the log. I do not think that pedagogy in the pre-Civil War period in the United States was very impressive, or after the Civil War for that matter—when a professor at Yale shot an undergraduate because he misbehaved. But is it now possible with a much more open kind of selection to produce that many teachers who will possess the capacity to stimulate and excite and who can convey the values and have the breadth which the most demanding students insist on or who will have the capacity for learning, hard work, imagination, and responsibility which we desire? Does the human race have enough of those capacities at a time when there are so many other great demands for talent in so many important fields of social, economic, political, and cultural life? The United States and other countries are increasingly undertaking more than they have the brains to do, more than they have the personality qualities to perform. So we stagger on as well as we can.

We have to listen to the rhetoric of these young people, but we also have to explain to them the determinants of the human situation. There are limits, unless you can change the genetic pool.

PIZZORNO

I should like to respond to the last exchange of comments. Before Mr. Meyerson spoke, I was uneasy because there was too much consensus in our group. Everything is not so easy as we are trying to picture it here. The problems are much deeper than shifting a bit from the research side to the teaching side. Asa Briggs commented that there are not really contradictions, but tensions. Thus we moved to the middle of the road and became very happy. I thought then that we had to see if there was not something challenging us a little more deeply than we had been willing to admit.

I will just try to take two points established in our consensus, and discuss the consequences of those two points. Everybody agreed that theoretical or philosophical teaching is more agreeable to the teacher. We seem to agree that the higher the theoretical content of the teaching, the greater the incentive for the professor to teach. I tried to link this point of consensus with the other point that came up early this morning about the erotic content of the communication between a teacher and a pupil. Let us call it emotional; it is not really erotic. Earlier the professor would have been on the sadistic side of this relationship; now he has been forced over to the masochistic side. We must somehow break the pattern. On this point I am not in agreement with Mr. Killy. Given what has happened in the last few years, we cannot now have this kind of personal relationship with students. As an element in the process of socialization today, the old Socratic relationship is too emotionally loaded. The relationship must become more formal, more sober. I do not agree with Professor Shils; students do resent this enforced paternalistic intimacy.

SHILS

You cannot take them at their word. They say one thing and mean another; they will even say different things in the course of a conversation.

PIZZORNO

Let me say, then, that *I* do not like to have this kind of relationship with my students. It is no longer a help to have this kind of relationship. We cannot think that we are able to socialize them by means of such a situation. *In loco parentis* is a thing of the past; we are no longer a father substitute. I personally do not like this kind of relationship with my students, nor do I think it a healthy one to have.

There should be some sort of bargaining position in our position with the students. Let us bargain, and let us give what we can in our teaching or our research. But we should not try to socialize them or to give them some sort of general moral introduction to life. They do not need this.

KILLY

The few erotic experiences I know of have never been paternalistic.

ROBINSON

I feel a need to respond to Edward Shils, and in support of Martin Meyerson. As Martin spoke, the specific items he mentioned elicited a general nodding. It was the larger and fuzzier issues about values that brought the disagreement. Professor Shils' basis retort was: Can we give it to them? But when people come to the university community and say they want to crack an atom, the first response is not essentially the negative one: Are they asking too much of us? In general, we tend to grasp at these ideas; we will go back and try. Rather than the defensiveness of Professor Shils' reply, it seems to me that our first response should be to try and find out the things we can do. The students are, after all, our basic customers; they are the people who will use the knowledge that we spend all of our time trying to uncover.

SHILS

Giving them knowledge that they can use is a different thing from what they are asking for. I do not want to overgeneralize, by any means, and say that all the students are like this, but the small minority that Martin was speaking for does not want knowledge or a technique of understanding; these students simply want to discomfort and disrupt institutions, and that has to be acknowledged. This is a small minority, but it can become quite significant.

ROBINSON

Some people want space exploration simply for militaristic purposes. We do not say, therefore, that we are not going to explore space.

BRIGGS

I think Professor Pizzorno was getting into extremely difficult territory when he implied that the relationship between the tutor and student is absolutely divorced from the structural state of the university in which the tutor and student operate. You can draw sharp distinctions between the fundamental structural problems within universities and the problems of the psychological relationships between tutors and students. In certain circumstances these relationships are impossible because of the nature of the structural setups. In this sense, we have been distinguishing too much, if you like, between the two halves of Marx—between the Marx who deals with individual relations and the Marx who deals with economic patterns of growth. We had a long discussion yesterday about patterns of resource allocation that are fundamental to the university problems in all our countries. This morning we have been discussing relationships of a psychological character between individuals and the terribly difficult problems of socialization. I do not think there are any easy answers, and I should like to experiment and to innovate, as Professor Crozier suggested. But we can separate the two discussions only if we bear in mind that we live in societies where universities have got to interpret themselves to the communities that are paying the money and where they also have got their own internal relationships to settle. To keep the two halves absolutely separate skews the picture.

All societies want scientific research, whatever their political structure. When we start talking about humanities, we get into problems of a quite different order. When Professor Pizzorno was talking, as he was in such an extremely austere way, about no possibilities of socialization, no kind of contact other than the straight communication of organizational knowledge, he was describing what to me would mean the end of the university as an institution, indeed almost the end of knowledge. Knowledge is culture, and if universities abdicate from that whole area, they are going to have a tough time indeed over the next hundred years.

PIZZORNO

You can have this socialization implicitly, but you cannot aim for it explicitly.

BOURRICAUD

I should like to put in the pot not an altogether new element, but something perhaps slightly different. I should like very much to see the crisis of the academic ethos discussed with a great deal more clarity. At the time of Durkheim, for instance, some sort of balance had been achieved between teaching and research, though the research was not of the contemporary type. For a certain set of reasons that balance has been lost, and a new situation has emerged. I would like to explore the causes for the *anomie* in which the academic profession finds itself in France today. There are two or three alienating mechanisms that would be interesting to take into account. Perhaps if we did consider these mechanisms, they would explain why one has at the same time in France a breakdown of two main components of the role (these being research and teaching) as well as a loss in status, prestige, and control by the profession. There is a great disjunction between the present state of knowledge within the profession and the type of course or teaching that is administered by the professor or teaching person in the university system. This gap has alienated the academic group from the most advanced and better informed among young people and those without the proper academic credentials. In France, the separation between the teaching people and the research people resulted to some extent from the unsatisfactory formation of the teachers and of the professors themselves.

Second, I should further like to analyze the mechanism of alienation of the teacher from the student body. In that respect, we would have to devote much attention to the lecture. It is remarkable to see how much time, for instance, a man like Durkheim devoted to writing his lectures. But of course, even in that case, this formalized rite was perhaps of more significance for the teacher or the professor than for the student body. On that score, I would agree with Lord Fulton in his previous intervention.

Third, we might perhaps explore the alienation of the university from any of the leading and deciding agencies, as it becomes insulated within its own corporate and traditional problems. My analysis is taken from the French case, which is a particular type of university pathology, but to some extent it could also apply to other cases. After that deviance mechanism has been analyzed, we might ask what sort of countermechanism could be set in motion so that we could start the operation recuperating.

Conference Participants

Conference on New Trends in History: May 23-25, 1968

Samuel H. Beer
Lee Benson
Felix Gilbert
Stephen R. Graubard
David J. Herlihy
Stanley Hoffmann
Carl Kaysen
Leonard Krieger
Thomas S. Kuhn

David Landes
Joseph Levenson
Frank E. Manuel
J. G. A. Pocock
David J. Rothman
Carl E. Schorske
Lawrence Stone
Charles Tilly

Conference on the Languages of the Humanistic Studies: May 4-7, 1968

Meyer Abrams
Antonio Alatorre
Morton W. Bloomfield
Peter J. Caws
Clifford Geertz
Stephen R. Graubard
Geoffrey Hartman
E. D. Hirsch
Murray Krieger
George Kubler
Martin Malia

William McGuire
J. Hillis Miller
Leonard Newmark
Talcott Parsons
Roy Harvey Pearce
Roger Revelle
Henry Nash Smith
Arnold Stein
Karl Uitti
Eric Weil

Conference on Governance of Universities: Rights and Responsibilities: April 5 and 6, 1968

Daniel Bell
Laura Bornholdt
John Brademas

Jill Conway
Martin Duberman
Robert J. Glaser

Stephen R. Graubard
Hanna H. Gray
Carl Kaysen
Edward H. Levi
Martin Meyerson
Robert S. Morison

Talcott Parsons
Bruce L. Payne
David Riesman
Neil R. Rudenstine
Preble Stolz
David M. Wax

Conference on Governance of Universities: Rights and Responsibilities: November 14-15, 1968

Peter H. Armacost
Robert F. Bacher
Louis T. Benezet
Landrum R. Bolling
Laura Bornholdt
Peter J. Caws
Jill Conway
Sarah E. Diamant
C. M. Dick, Jr.
Ralph A. Dungan
Seymour Eskow
Edgar Z. Friedenberg
Stephen R. Graubard
Hanna H. Gray
Andrew M. Greeley
Jeff Greenfield
Eugene E. Grollmes
Vivian W. Henderson
Gerald Holton
Willard Hurst
Dexter M. Keezer
Clark Kerr
Gardner Lindzey
S. E. Luria
David E. Matz
Jean Mayer

Walter P. Metzger
Martin Meyerson
Robert S. Morison
Henry Norr
Talcott Parsons
Bruce L. Payne
J. W. Peltason
Harvey Picker
Roger Revelle
Leonard M. Rieser
Philip C. Ritterbush
Olin C. Robinson
Neil R. Rudenstine
Elizabeth Sewell
William H. Sewell
Edward Joseph Shoben, Jr.
John R. Silber
Charles E. Silberman
R. L. Sproull
Preble Stolz
Kenneth S. Tollett
Martin Trow
Heinz Von Foerster
George R. Waggoner
Andrew T. Weil
W. Max Wise

Conference on Higher Education in Industrial Societies: November 29 and 30, 1968

Joseph Ben-David
François Bourricaud

Asa Briggs
Mauro Cappelletti

Conference Participants

Michel Crozier
A. D. Dunton
Baron John Scott Fulton
Pierre Grappin
Stephen R. Graubard
Walther Killy
Max Kohnstamm
Yoichi Maeda
Pierre Massé

Martin Meyerson
Elting E. Morison
Victor G. Onushkin
Talcott Parsons
Alessandro Pizzorno
Marshall Robinson
Edward Shils
Sir Solly Zuckerman

Notes on Contributors

Samuel H. Beer
Professor of Government
Harvard University

Daniel Bell
Professor of Sociology
Harvard University

Joseph Ben-David
Professor of Sociology
Hebrew University of Jerusalem

Lee Benson
Professor of History
University of Pennsylvania

Landrum R. Bolling
President
Earlham College

François Bourricaud
Professor of Sociology
The Sorbonne
University of Paris

John Brademas
United States House
 of Representatives

Asa Briggs
Vice Chancellor and
 Professor of History
Sussex University

Mauro Cappelletti
Professor of Law and
 Director, Institute of
 Comparative Law
University of Florence

Peter J. Caws
Professor of Philosophy
Hunter College

Jill Conway
Assistant Professor of History
University of Toronto

Michel Crozier
Maître de Recherche au Centre
 National de la Recherche
 Scientifique
Director, Centre de Sociologie des
 Organisations
Paris, France

Sarah E. Diamant
Doctoral Candidate in American
 History
Cornell University

C. M. Dick, Jr.
President
Business Equipment Manufacturers
 Association

Martin Duberman
Professor of History
Princeton University

A. D. Dunton
President
Carleton University
Ottawa, Canada

Seymour Eskow
President
Rockland Community College

Edgar Z. Friedenberg
Professor of Education and
 Sociology
State University of New York
 at Buffalo

Notes on Contributors

Baron John Scott Fulton
Chairman
The British Council

Clifford Geertz
Professor of Anthropology
University of Chicago

Felix Gilbert
School of Historical Studies
Institute for Advanced Study

Robert J. Glaser
Dean and Professor of Medicine
Medical School
Stanford University

Pierre Grappin
Professor of German Literature
    and History
University of Paris (Nanterre)

Stephen R. Graubard
Professor of History
Brown University

Hanna H. Gray
Associate Professor of History
University of Chicago

Andrew M. Greeley
Program Director
National Opinion Research Center

Jeff Greenfield
Assistant to the Mayor
New York City

Eugene E. Grollmes
Administrative Intern
St. Louis University

David J. Herlihy
Professor of History
University of Wisconsin

Stanley Hoffmann
Professor of Government
Harvard University

Gerald Holton
Professor of Physics
Harvard University

Willard Hurst
Villas Professor of Law
University of Wisconsin

Carl Kaysen
Director
Institute for Advanced Study

Dexter M. Keezer
Economic Adviser
McGraw-Hill, Incorporated

Clark Kerr
Chairman and Executive Director
Carnegie Commission on Higher
    Education

Walther Killy
Professor of Literature
University of Göttingen

Leonard Krieger
University Professor of History
University of Chicago

Thomas S. Kuhn
M. Taylor Pyne Professor of the
    History of Science
Princeton University

David Landes
Professor of History
Harvard University

Joseph Levenson
deceased, formerly
Sather Professor of History
University of California at Berkeley

Edward H. Levi
President
University of Chicago

S. E. Luria
Sedgwick Professor of Biology
The Massachusetts Institute of
    Technology

Yoichi Maeda
Professor of French Language
    and Culture
University of Tokyo

1229

Martin Malia
Professor of History
University of California at Berkeley

Frank E. Manuel
Professor of History
New York University

Jean Mayer
Professor of Nutrition
Harvard University

Walter P. Metzger
Professor of History
Columbia University

Martin Meyerson
President
State University of New York at Buffalo

Elting E. Morison
Master, Timothy Dwight College
and Professor of History
Yale University

Robert S. Morison
Director
Division of Biological Sciences
Cornell University

Henry Norr
Graduate Student
School of Education
Harvard University

Victor G. Onushkin
Senior Staff Member
International Institute for Educational Planning
UNESCO (Paris)

Talcott Parsons
Professor of Social Relations
Harvard University

Bruce L. Payne
Instructor of Government
Kirkland College

Alessandro Pizzorno
Professor of Sociology
University of Urbino

J. G. A. Pocock
Professor of History and Political Science
Washington University

Roger Revelle
Director
Center for Population Studies
Harvard University

David Riesman
Henry Ford II Professor of Social Sciences
Harvard University

Philip C. Ritterbush
Director
Office of Academic Programs
Smithsonian Institution

Marshall Robinson
Program Officer in Charge
Division of Education and Research
The Ford Foundation

David J. Rothman
Associate Professor of History
Columbia University

Neil R. Rudenstine
Associate Professor of English and Dean of Students
Princeton University

Carl E. Schorske
Professor of History
Princeton University

Edward Shils
Committee on Social Thought
University of Chicago

Edward Joseph Shoben, Jr.
Chairman, Council on Higher Education Studies, and Director, Center for Higher Education
State University of New York at Buffalo

John R. Silber
Professor of Philosophy and Dean of the College of Arts and Sciences
University of Texas at Austin

Notes on Contributors

Charles E. Silberman
Director
The Carnegie Study of the Education of Educators

R. L. Sproull
Vice President and Provost
The University of Rochester

Preble Stolz
Professor of Law
University of California at Berkeley

Lawrence Stone
Dodge Professor of History and Director of the Shelby Cullom Davis Center for Historical Studies
Princeton University

Charles Tilly
Professor of Sociology and History
University of Michigan

Kenneth S. Tollett
Dean
School of Law
Texas Southern University

George R. Waggoner
Dean, College of Liberal Arts and Sciences, and Professor of English
University of Kansas

David M. Wax
Assistant to the Administrator
Housing and Development Administration
New York City

Eric Weil
Professor of Philosophy
University of Nice

Sir Solly Zuckerman
Chief Scientific Advisor
Cabinet Officer
London, England

DÆDALUS

INDEX
1969

DÆDALUS
Journal of the American Academy of Arts and Sciences

VOLUME 98

of the Proceedings of the Academy

1969

BOARD OF EDITORS

Raymond Aron, John E. Burchard, Clarence Faust, Karl von Frisch, Etienne Gilson, Gerald Holton, Clark Kerr, R. M. MacIver, Archibald MacLeish, Talcott Parsons, Paul B. Sears, Harlow Shapley, George Wald

Stephen R. Graubard, *Editor of the Academy and of Dædalus;* Geno A. Ballotti, *Managing Editor;* Judith Williams, *Manuscripts Editor*

Editorial Associates: Judith E. Boone, Patricia Cumming, Bonnie J. Harris, Ena Lubicz-Nycz

EDITORIAL OFFICE: *Dædalus,* the Journal of the American Academy of Arts and Sciences, 7 Linden Street, Harvard University, Cambridge, Massachusetts 02138.

SUBSCRIPTION OFFICE: American Academy of Arts and Sciences, 280 Newton Street, Brookline Station, Boston, Massachusetts 02146.

Published by the American Academy of Arts and Sciences.

Printed in the United States of America. Printing Office: 2901 Byrdhill Road, Richmond, Virginia. Library of Congress Catalog Card Number: 12-30299. Second-class postage paid at Boston, Massachusetts, and at additional mailing office. Copyright © 1969 by the American Academy of Arts and Sciences.

CONTENTS OF DÆDALUS, VOLUME 98, 1969

No. 1, Winter 1969, "Perspective on Business"

	v	Preface
WILLIAM LETWIN	1	The Past and Future of the American Businessman
ALFRED D. CHANDLER, JR.	23	The Role of Business in the United States: A Historical Survey
RICHARD H. HOLTON	41	Business and Government
ROBERT T. AVERITT	60	American Business: Achievement and Challenge
ELI GOLDSTON	78	New Prospects for American Business
RAYMOND VERNON	113	The Role of U. S. Enterprise Abroad
NEIL W. CHAMBERLAIN	134	The Life of the Mind in the Firm
MICHEL CROZIER	147	A New Rationale for American Business
R. JOSEPH MONSEN	159	The American Business View
LEONARD S. SILK	174	Business Power, Today and Tomorrow
ANDREW SHONFIELD	191	Business in the Twenty-First Century
	208	Notes on Contributors
	210	Recent Issues of Dædalus

ADVISORY COMMITTEE

Kenneth R. Andrews, Daniel Bell, Warren Bennis, L. Earle Birdzell, Jr., Wayne G. Broehl, Harvey Brooks, Neil W. Chamberlain, Alfred D. Chandler, Jr., M. B. E. Clarkson, Thomas C. Cochran, Michel Crozier, Anthony Downs, Henry W. Ehrmann, Gaylord A. Freeman, Jr., Thomas M. Garrett, Alexander Gerschenkron, Eli Goldston, John W. Hennessey, Jr., Karl Hill, Richard H. Holton, Randall T. Klemme, William Letwin, Seymour Martin Lipset, Robert M. Macdonald, Daniel Marx, Jr., Isamu Miyazaki, R. Joseph Monsen, Herbert C. Morton, Bernard J. Muller-Thym, John C. Parkin, Theodore V. Purcell, David Riesman, John H. Rubel, Andrew Shonfield, Leonard S. Silk, Raymond Stevens, Raymond Vernon, E. O. Vetter

No. 2, Spring 1969, "Ethical Aspects of Experimentation with Human Subjects"

	v	Preface
PAUL A. FREUND	viii	Introduction to the Issue "Ethical Aspects of Experimentation with Human Subjects"
HANS JONAS	219	Philosophical Reflections on Experimenting with Human Subjects
HERRMAN L. BLUMGART	248	The Medical Framework for Viewing the Problem of Human Experimentation
HENRY K. BEECHER	275	Scarce Resources and Medical Advancement
PAUL A. FREUND	314	Legal Frameworks for Human Experimentation
TALCOTT PARSONS	325	Research with Human Subjects and the "Professional Complex"
MARGARET MEAD	361	Research with Human Beings: A Model Derived from Anthropological Field Practice
GUIDO CALABRESI	387	Reflections on Medical Experimentation in Humans
LOUIS L. JAFFE	406	Law as a System of Control
DAVID F. CAVERS	427	The Legal Control of the Clinical Investigation of Drugs: Some Political, Economic, and Social Questions
LOUIS LASAGNA	449	Special Subjects in Human Experimentation
GEOFFREY EDSALL	463	A Positive Approach to the Problem of Human Experimentation
JAY KATZ	480	The Education of the Physician-Investigator
FRANCIS D. MOORE	502	Therapeutic Innovation: Ethical Boundaries in the Initial Clinical Trials of New Drugs and Surgical Procedures
DAVID D. RUTSTEIN	523	The Ethical Design of Human Experiments

WILLIAM J. CURRAN	542	Governmental Regulation of the Use of Human Subjects in Medical Research: The Approach of Two Federal Agencies
	595	Advisory Committee
	596	Notes on Contributors
	598	Recent Issues of Dædalus

ADVISORY COMMITTEE

John D. Arnold, Bernard Barber, Lewis W. Beck, Henry K. Beecher, Nathaniel I. Berlin, Alexander M. Bickel, Herrman L. Blumgart, Orville G. Brim, Jr., Guido Calabresi, William M. Carley, David F. Cavers, Alexander L. Clark, Eugene A. Confrey, Theodore Cooper, William J. Curran, Arthur Dyck, Geoffrey Edsall, Renée C. Fox, Paul A. Freund, James L. Goddard, Erving Goffman, Hudson Hoagland, Louis L. Jaffe, Hans Jonas, Jay Katz, John H. Knowles, Irving Ladimer, Thomas Langan, Louis Lasagna, Walsh McDermott, Ronald W. McNeur, Margaret Mead, Everett Mendelsohn, Francis D. Moore, Talcott Parsons, Theodore V. Purcell, Gardner C. Quarton, Robert L. Ringler, Roy E. Ritts, Jr., Oscar M. Ruebhausen, Paul S. Russell, David D. Rutstein, Blair L. Sadler, George E. Schreiner, William B. Schwartz, Majorie P. Wilson

No. 3, Summer 1969, "The Future of the Humanities"

	v	Preface
JAMES S. ACKERMAN	605	Introduction to the Issue "The Future of the Humanities"
		I. General
WALTER J. ONG	617	Crisis and Understanding in the Humanities
G. JON ROUSH	641	What Will Become of the Past?
HERBERT BLAU	654	Relevance: The Shadow of a Magnitude
ROGER W. SHATTUCK	677	Thoughts on the Humanities
ROBERT COLES	684	The Words and Music of Social Change
R. J. KAUFMANN	699	On Knowing One's Place: A Humanistic Meditation
	715	Excerpts from Conference I
		II. The Arts
LEON KIRCHNER	739	Notes on Understanding
WOLF KAHN	747	Uses of Painting Today

JESSE REICHEK	755	A Painter and Teacher as Amateur Humanist

Illustrations

JESSE REICHEK	Drawing 1968
ERIC MARTIN	Light Pen Sequence Sketch, 1969
LEON KIRCHNER	"Quartet #3"
WOLF KAHN	"Declivity with Apple Trees," Drawing 1968

MICHAEL O'HARE	765	Designers' Dilemma
	779	Excerpts from Conference II

III. The Disciplines

GERALD F. ELSE	803	The Old and the New Humanities
CEDRIC H. WHITMAN	809	Why Not the Classics? An Old-Fashioned View
MICHAEL C. J. PUTNAM	815	Notes on Classical Studies
LEO STEINBERG	824	Objectivity and the Shrinking Self
JOHN M. ROSENFIELD	837	The Arts in the Realm of Values
LEO TREITLER	844	On Responsibility and Relevance in Humanistic Disciplines

IV. Conclusion

JAMES S. ACKERMAN	855	Two Styles: A Challenge to Higher Education
	870	Conference Participants
	871	Notes on Contributors
	875	Recent Issues of Dædalus

ADVISORY COMMITTEE

James S. Ackerman, Stanford O. Anderson, William Arrowsmith, Herbert Blau, Lawrence W. Chisolm, Marshall Cohen, Tom Cole, Robert Coles, Edward F. D'Arms, Richard M. Douglas, Gerald F. Else, John Higham, John Holt, Wolf Kahn, R. J. Kaufmann, Barnaby Keeney, Joseph W. Kerman, Leon Kirchner, Eric Martin, Leo Marx, Ernst Mayr, Leonard B. Meyer, George W. Morgan, Charles Muscatine, Arnold S. Nash, Howard Nemerov, Michael O'Hare, Richard Ohmann, Walter J. Ong, Franklin Patterson, R. L. Predmore, Michael C. J. Putnam, Jesse Reichek, John M. Rosenfield, G. Jon Roush, Roger W. Shattuck, Leo Steinberg, George Steiner, Charles L. Stevenson, Walter Sutton, Leo Treitler, Herbert Weisinger, Lynn T. White, Jr., Cedric H. Whitman

No. 4, Fall 1969, "Dialogues"

- *v* *Preface*
- *888* *New Trends in History*
- *978* *The Languages of the Humanistic Studies*
- *1030* *Governance of the Universities I*
- *1092* *Governance of the Universities II*
- *1157* *Higher Education in Industrial Societies*
- *1225* *Conference Participants*
- *1228* *Notes on Contributors*
- *1233* *Index of Authors of Dædalus, Volume 98, 1969*
- *1244* *Recent Issues of Dædalus*

Index of Authors, *DÆDALUS*, Volume 98, 1969

Ackerman, James S. 605, 855
Averitt, Robert T. 60

Beecher, Henry K. 275
Blau, Herbert 654
Blumgart, Herrman L. 248

Calabresi, Guido 387
Cavers, David F. 427
Chamberlain, Neil W. 134
Chandler, Alfred D., Jr. 23
Coles, Robert 684
Crozier, Michel 147
Curran, William J. 542

Edsall, Geoffrey 463
Else, Gerald F. 803

Freund, Paul A. viii (No. 2), 314

Goldston, Eli 78

Holton, Richard H. 41

Jaffe, Louis L. 406
Jonas, Hans 219

Kahn, Wolf 747, illustration (No. 3)
Katz, Jay 480
Kaufmann, R. J. 699

Kirchner, Leon 739, illustration (No. 3)

Lasagna, Louis 449
Letwin, William 1

Martin, Eric illustration (No. 3)
Mead, Margaret 361
Monsen, R. Joseph 159
Moore, Francis D. 502

O'Hare, Michael 765
Ong, Walter J. 617

Parsons, Talcott 325
Putnam, Michael C. J. 815

Reichek, Jesse 755, illustration (No. 3)
Rosenfield, John M. 837
Roush, G. Jon 641
Rutstein, David D. 523

Shattuck, Roger 677
Shonfield, Andrew 191
Silk, Leonard 174
Steinberg, Leo 824

Treitler, Leo 844

Vernon, Raymond 113

Whitman, Cedric H. 809

Index of Transcript Remarks, *DÆDALUS*, Volume 98, 1969

No. 3, Summer 1969, "The Future of the Humanities"

Ackerman, James S. 719, 720, 722, 723
Anderson, Stanford O. 721
Arrowsmith, William 724, 732

Blau, Herbert 729, 730, 733, 780, 781, 786, 794

Chisolm, Lawrence W. 782, 784
Cohen, Marshall 723, 731, 732
Coles, Robert 787, 788, 798, 800

D'Arms, Edward F. 786, 787

Graubard, Stephen R. 796, 799, 800

Holt, John 733

Kahn, Wolf 790, 791
Kaufmann, R. J. 784, 785, 795
Kirchner, Leon 788, 789, 790

Marx, Leo 715, 716, 717, 727, 779, 780, 781, 782, 783, 784, 787, 793, 795, 796, 798, 799
Meyer, Leonard B. 786, 793, 794
Morgan, George W. 726
Muscatine, Charles 717, 718, 719, 724, 734

O'Hare, Michael 783, 785, 786, 794, 798, 799

Ohmann, Richard 721, 722, 797, 798
Predmore, R. L. 726
Roush, G. Jon 797
Steinberg, Leo 727, 728, 729, 791, 792, 793, 796, 797
Steiner, George 724, 725, 726, 732, 733, 734
Stevenson, Charles L. 782, 783
Weisinger, Herbert 785, 798
White, Lynn T., Jr. 730, 731, 734, 735

No. 4, Fall 1969, "Dialogues"

Beer, Samuel H. 909, 920, 921, 926, 933, 950, 951, 952
Bell, Daniel 1033, 1034, 1035, 1036, 1040, 1041, 1043, 1046, 1051, 1060, 1072, 1082, 1088, 1089, 1090, 1091
Ben-David, Joseph 1166, 1167, 1168, 1183, 1186, 1212, 1213
Benson, Lee 891, 892, 901, 902, 904, 905, 910, 939, 940, 941, 942, 943, 945, 953, 954, 955, 959
Bolling, Landrum R. 1107, 1108, 1115, 1116
Bourricaud, François 1190, 1191, 1222, 1223
Brademas, John 1084, 1085, 1086, 1087, 1088, 1089, 1090
Briggs, Asa 1164, 1165, 1172, 1178, 1179, 1180, 1181, 1200, 1210, 1211, 1212, 1221, 1222

Cappelletti, Mauro 1185
Caws, Peter J. 1114, 1115, 1116, 1117, 1118, 1124, 1126, 1127
Conway, Jill 1049, 1050, 1058, 1059, 1077, 1078
Crozier, Michel 1165, 1180, 1181, 1196, 1197, 1210, 1216

Diamant, Sarah E. 1148, 1149
Dick, C. M., Jr. 1103, 1104
Duberman, Martin 1036, 1045, 1052, 1053, 1072, 1074, 1075, 1077, 1083
Dunton, A. D. 1173, 1174, 1215, 1216

Eskow, Seymour 1123, 1124

Friedenberg, Edgar Z. 1105, 1106
Fulton, Baron John Scott 1214, 1215

Geertz, Clifford 981-992
Gilbert, Felix 891, 897, 898, 908, 912, 923, 929, 937, 949, 950, 955, 956, 963, 966, 968, 975

Glaser, Robert J. 1065, 1073, 1074, 1085
Grappin, Pierre 1193, 1194, 1195, 1196, 1197, 1198, 1199, 1200, 1201
Graubard, Stephen R. 905, 934, 935, 951, 952, 954, 955, 969, 970, 1042, 1043, 1138, 1162, 1163, 1190, 1191
Gray, Hanna H. 1056, 1057, 1058, 1067, 1078, 1079
Greeley, Andrew M. 1133
Greenfield, Jeff 1107, 1108, 1109, 1116, 1117, 1118, 1121, 1122, 1123, 1124
Grollmes, Eugene E. 1110, 1111

Herlihy, David J. 937, 938, 939, 940, 964, 965
Hoffmann, Stanley 910, 911, 919, 920, 921, 927, 948, 949, 952, 953, 954, 959, 965, 970, 971, 972, 973
Holton, Gerald 1103, 1106, 1118, 1119, 1127, 1128
Hurst, Willard 1152, 1153

Kaysen, Carl 902, 904, 905, 908, 909, 926, 932, 938, 939, 940, 941, 943, 959, 966, 968, 969, 973, 1036, 1037, 1038, 1039, 1043, 1044, 1051, 1052, 1061, 1062, 1063, 1064, 1066, 1067, 1070, 1075, 1076, 1077, 1082, 1086, 1087, 1088
Keezer, Dexter M. 1134
Kerr, Clark 1104, 1119, 1120, 1126
Killy, Walther 1188, 1199, 1200, 1213, 1214, 1220
Krieger, Leonard 911, 912, 923, 935, 966, 974
Kuhn, Thomas S. 896, 897, 928, 943, 944, 969, 971, 972, 973, 975, 976

Landes, David 897, 898, 914, 915, 927, 944, 945, 951, 952, 958, 963, 964, 966, 967, 971, 972, 974

Levenson, Joseph 903, 904, 915, 916, 917, 918, 947, 948, 961, 962, 963, 967, 968, 969, 975
Levi, Edward H. 1046, 1047, 1063, 1070, 1071, 1079, 1080
Luria, S. E. 1109, 1110, 1114

Maeda, Yoichi 1216, 1217
Malia, Martin 1019-1028
Manuel, Frank E. 892, 893, 894, 899, 900, 913, 914, 915, 916, 924, 925, 926, 927, 928, 929, 937, 945, 946, 947, 959, 960, 974
Mayer, Jean 1130, 1131
Metzger, Walter P. 1133, 1134, 1138, 1139, 1140, 1141, 1142, 1143, 1144, 1145, 1146, 1147, 1148
Meyerson, Martin 1039, 1040, 1048, 1049, 1075, 1084, 1089, 1090, 1113, 1128, 1129, 1130, 1131, 1132, 1133, 1135, 1136, 1151, 1152, 1165, 1166, 1168, 1171, 1175, 1176, 1177, 1178, 1181, 1182, 1203, 1204, 1208, 1209, 1217, 1218
Morison, Elting E. 1171, 1172, 1185, 1186, 1201, 1202, 1203, 1209, 1210, 1213, 1216
Morison, Robert S. 1047, 1048, 1067, 1068, 1069, 1080, 1081, 1103, 1108, 1113, 1114, 1117, 1118, 1125, 1126, 1127

Norr, Henry 1111

Onushkin, Victor G. 1186, 1187, 1191, 1192, 1193

Parsons, Talcott 993-1000; 1039, 1045, 1053, 1054, 1062, 1063, 1064, 1069, 1070, 1074, 1088, 1105, 1111, 1112, 1163, 1164, 1177, 1184, 1185, 1207, 1208, 1209, 1214
Payne, Bruce L. 1044, 1055, 1056, 1059, 1060, 1064, 1065, 1067, 1068, 1073, 1074, 1075, 1076, 1077, 1081, 1082, 1083, 1136, 1153, 1154
Pizzorno, Alessandro 1172, 1173, 1188, 1189, 1198, 1206, 1219, 1220, 1222
Pocock, J. G. A. 895, 896, 898, 899, 901, 902, 906, 907, 908, 909, 911, 916, 917, 919, 920, 921, 922, 923, 929, 932, 942, 943, 951, 965, 967, 968

Revelle, Roger 1007-1018; 1136, 1137
Riesman, David 1034, 1037, 1041, 1042, 1054, 1055, 1058, 1061, 1065, 1066, 1067, 1068, 1074, 1083, 1085
Ritterbush, Philip C. 1106, 1107, 1126
Robinson, Marshall 1174, 1175, 1220, 1221
Rothman, David J. 894, 895, 905, 906, 912, 913, 915, 935, 940, 947, 954, 965, 970, 975
Rudenstine, Neil R. 1050, 1051, 1063, 1064, 1079, 1080, 1109

Schorske, Carl E. 898, 904, 912, 926, 927, 929, 930, 931, 932, 933, 934, 935, 936, 937, 946, 950, 955, 958, 959, 961, 963, 967, 968, 973, 974, 975
Shils, Edward 1183, 1184, 1189, 1190, 1206, 1207, 1208, 1212, 1217, 1218, 1219, 1220, 1221
Shoben, Edward Joseph, Jr. 1120, 1121, 1122
Silber, John R. 1132, 1149, 1150, 1151
Silberman, Charles E. 1104, 1124, 1125
Sproull, R. L. 1137
Stolz, Preble 1060, 1061
Stone, Lawrence 894, 900, 901, 914, 942, 956, 957, 958, 971

Tilly, Charles 922, 936, 944, 945, 946, 947, 948, 949, 950, 960
Tollett, Kenneth S. 1123

Waggoner, George R. 1112, 1113, 1114
Wax, David M. 1069
Weil, Eric 1001-1006

Zuckerman, Sir Solly 1164, 1168, 1169, 1170, 1171, 1177, 1178, 1199, 1200, 1201, 1204, 1205

Issues of *DÆDALUS* in Print

These may be ordered from the *Dædalus* Subscription Office, the American Academy of Arts and Sciences, 280 Newton Street, Brookline Station, Boston, Massachusetts 02146.

THE FUTURE OF THE HUMANITIES *(Summer 1969)*

ETHICAL ASPECTS OF EXPERIMENTATION WITH HUMAN SUBJECTS *(Spring 1969)*

PERSPECTIVES ON BUSINESS *(Winter 1969)*

THE CONSCIENCE OF THE CITY *(Fall 1968)*

PHILOSOPHERS AND KINGS: STUDIES IN LEADERSHIP *(Summer 1968)*

HISTORICAL POPULATION STUDIES *(Spring 1968)*

STUDENTS AND POLITICS *(Winter 1968)*

AMERICA'S CHANGING ENVIRONMENT *(Fall 1967)*

TOWARD THE YEAR 2000: WORK IN PROGRESS *(Summer 1967)*

COLOR AND RACE *(Spring 1967)*

RELIGION IN AMERICA *(Winter 1967)*

FICTION IN SEVERAL LANGUAGES *(Fall 1966)*

TRADITION AND CHANGE *(Summer 1966)*

CONDITIONS OF WORLD ORDER *(Spring 1966)*

THE NEGRO AMERICAN—2 (special issue, *Winter 1966*)

THE NEGRO AMERICAN—1 (special issue, *Fall 1965*)

CREATIVITY AND LEARNING *(Summer 1965)*

UTOPIA *(Spring 1965)*

SCIENCE AND CULTURE *(Winter 1965)*

THE CONTEMPORARY UNIVERSITY: U.S.A. *(Fall 1964)*

POPULATION, PREDICTION, CONFLICT, EXISTENTIALISM *(Summer 1964)*

Continued

THE PROFESSIONS *(Fall 1963)*

THEMES IN TRANSITION *(Summer 1963)*

PERSPECTIVES ON THE NOVEL *(Spring 1963)*

THE AMERICAN READING PUBLIC *(Winter 1963)*

AMERICAN FOREIGN POLICY—FREEDOMS AND RESTRAINTS *(Fall 1962)*

CURRENT WORK AND CONTROVERSIES—2 *(Summer 1962)*

SCIENCE AND TECHNOLOGY IN CONTEMPORARY SOCIETY *(Spring 1962)*

EXCELLENCE AND LEADERSHIP IN A DEMOCRACY *(Fall 1961)*

MASS CULTURE AND MASS MEDIA *(Spring 1960)*

TENTH ANNIVERSARY INDEX: 1958–68

ISSUES OF *DÆDALUS* OUT OF PRINT

See hard-cover books based on these issues, listed on following pages.

THE WOMAN IN AMERICA *(Spring 1964)*

A NEW EUROPE? (special issue, *Winter 1964*)

YOUTH: CHANGE AND CHALLENGE *(Winter 1962)*

EVOLUTION AND MAN'S PROGRESS *(Summer 1961)*

ETHNIC GROUPS IN AMERICAN LIFE *(Spring 1961)*

THE FUTURE METROPOLIS *(Winter 1961)*

ARMS CONTROL (special issue, *Fall 1960*)

THE RUSSIAN INTELLIGENTSIA *(Summer 1960)*

THE VISUAL ARTS TODAY (special issue, *Winter 1960*)

QUANTITY AND QUALITY *(Fall 1959)*

CURRENT WORK AND CONTROVERSIES—1 *(Summer 1959)*

MYTH AND MYTHMAKING *(Spring 1959)*

EDUCATION IN THE AGE OF SCIENCE *(Winter 1959)*

ON EVIDENCE AND INFERENCE *(Fall 1958)*

THE AMERICAN NATIONAL STYLE *(Spring 1958)*

SCIENCE AND THE MODERN WORLD VIEW *(Winter 1958)*

BOOKS PUBLISHED FROM ISSUES OF *DÆDALUS*

Virtually all the issues of *Dædalus*, in expanded form, appear also as books in hard covers and in subsequent paperback editions. Inquiries should be directed to the respective publishers.

HARD-COVER EDITIONS

COLOR AND RACE, edited by John Hope Franklin. Houghton Mifflin Company, 1968. $6.95.

TOWARD THE YEAR 2000: WORK IN PROGRESS, edited by Daniel Bell. Houghton Mifflin Company, 1968. $6.50.

CONDITIONS OF WORLD ORDER, edited by Stanley Hoffmann. Houghton Mifflin Company, 1968. $6.50.

FICTION IN SEVERAL LANGUAGES, edited by Henri Peyre. Houghton Mifflin Company, 1968. $6.00.

CREATIVITY AND LEARNING, edited by Jerome Kagan. Houghton Mifflin Company, 1967. $6.95.

THE NEGRO AMERICAN, edited by Talcott Parsons and Kenneth B. Clark. Houghton Mifflin Company, 1966. $9.50.

UTOPIAS AND UTOPIAN THOUGHT, edited by Frank E. Manuel. Houghton Mifflin Company, 1966. $6.50.

THE CONTEMPORARY UNIVERSITY: U.S.A., edited by Robert S. Morison. Houghton Mifflin Company, 1966. $6.50.

SCIENCE AND CULTURE, edited by Gerald Holton. Houghton Mifflin Company, 1965. $6.00.

THE PROFESSIONS IN AMERICA, edited by Kenneth S. Lynn. Houghton Mifflin Company, 1965. $5.00.

THE WOMAN IN AMERICA, edited by Robert Jay Lifton. Houghton Mifflin Company, 1965. $6.00.

A NEW EUROPE?, edited by Stephen R. Graubard. Houghton Mifflin Company, 1964. $8.50.

THE AMERICAN READING PUBLIC, edited by Roger H. Smith. R. R. Bowker Company, 1964. $7.95.

Continued

YOUTH: CHANGE AND CHALLENGE, edited by Erik H. Erikson. Basic Books, 1963. $6.50.

EXCELLENCE AND LEADERSHIP IN A DEMOCRACY, edited by Stephen R. Graubard and Gerald Holton. Columbia University Press, 1962. $5.00.

EVOLUTION AND MAN'S PROGRESS, edited by Hudson Hoagland and Ralph W. Burhoe. Columbia University Press, 1962. $4.50.

THE FUTURE METROPOLIS, edited by Lloyd Rodwin. George Braziller, Inc., 1961. $5.00.

ARMS CONTROL, DISARMAMENT, AND NATIONAL SECURITY, edited by Donald G. Brennan. George Braziller, Inc., 1961. $6.00.

THE RUSSIAN INTELLIGENTSIA, edited by Richard Pipes. Columbia University Press, 1961. $4.50.

CULTURE FOR THE MILLIONS?, edited by Norman Jacobs. D. Van Nostrand Company, Inc., 1961. $4.95.

THE VISUAL ARTS TODAY, edited by Gyorgy Kepes. Wesleyan University Press, 1960. $8.50.

QUANTITY AND QUALITY, edited by Daniel Lerner. The Free Press of Glencoe, Inc., 1961. $4.50.

MYTH AND MYTHMAKING, edited by Henry A. Murray. George Braziller, Inc., 1960. $6.00.

EDUCATION IN THE AGE OF SCIENCE, edited by Brand Blanshard. Basic Books, 1959. $4.50.

EVIDENCE AND INFERENCE, edited by Daniel Lerner. The Free Press of Glencoe, Inc., 1959. $4.00.

SYMBOLISM IN RELIGION AND LITERATURE, edited by Rollo May. George Braziller, Inc., 1960. $5.00.

THE AMERICAN STYLE, edited by Elting E. Morison. Harper and Brothers, 1958. $5.00.

SCIENCE AND THE MODERN MIND, edited by Gerald Holton. Beacon Press, 1958. $3.95.

Continued

PAPERBACK EDITIONS

TOWARD THE YEAR 2000: WORK IN PROGRESS, edited by Daniel Bell. Beacon Press, 1969. $2.95.

FICTION IN SEVERAL LANGUAGES, edited by Henri Peyre. Beacon Press, 1969. $2.95.

RELIGION IN AMERICA, edited by Robert N. Bellah and William G. McLoughlin. Beacon Press, 1968. $3.45.

MYTH AND MYTHMAKING, edited by Henry A. Murray. Beacon Press, 1968. $2.95.

THE NEGRO AMERICAN, edited by Talcott Parsons and Kenneth B. Clark. Beacon Press, 1967. $3.95.

CREATIVITY AND LEARNING, edited by Jerome Kagan. Beacon Press, 1967. $2.45.

UTOPIAS AND UTOPIAN THOUGHT, edited by Frank E. Manuel. Beacon Press, 1967. $2.45.

SCIENCE & CULTURE, edited by Gerald Holton. Beacon Press, 1967. $2.45.

THE CONTEMPORARY UNIVERSITY: U.S.A., edited by Robert S. Morison. Beacon Press, 1967. $2.45.

THE WOMAN IN AMERICA, edited by Robert Jay Lifton. Beacon Press, 1967. $2.45.

A NEW EUROPE?, edited by Stephen R. Graubard. Beacon Press, 1967. $3.45.

THE PROFESSIONS IN AMERICA, edited by Kenneth S. Lynn. Beacon Press, 1967. $1.95.

THE CHALLENGE OF YOUTH, edited by Erik H. Erikson. Anchor Books, 1965. $1.45.

CULTURE FOR THE MILLIONS?, edited by Norman Jacobs. Beacon Press, 1965. $1.95.

LUZERNE COUNTY
COMMUNITY COLLEGE